PLC Controls with Structured Text (ST)

PREFACE

When I started as Assistant Professor (teacher) at Dania Academy, Randers, Denmark, one of my first tasks was to find books which were suited for the AP Graduate in Automation Engineering education. Especially within programming in Structured Text (ST) no relevant books exist. Books of about 300 pages only contained a few pages directly about ST, and only at a theoretical level.

My students demanded examples, guides and methods for ST programming. As a consequence, in January 2017, I started writing a material which was named:

"Get started with Structured Text"

Since then, the material has continuously been updated, extended and used in my lectures. The material has been highly demanded among my students and now I have finished this book, so other interested readers can gain knowledge by studying this book.

It is my hope that you will enjoy this book.

I shall thank my students, fellow teachers and colleagues for feedback and inspiration.

Comments, complaints, compliments and suggestions as to improvement are received positively.
Please, send them to TomMejerAntonsen@gmail.com

First edition issued June 2018

Please enjoy!

Tom Mejer Antonsen
Denmark, Randers, march 2019

Tom Mejer Antonsen

PLC Controls with Structured Text (ST)

IEC 61131-3 and best practice ST programming

© 2019 Tom Mejer Antonsen

2. Edition, March 2019

Illustrations and graphics: **Tom Mejer Antonsen**

Translated by: **Niels Kampmann**

The original version (Danish) 1. Edition issued March 2018

Publisher: Books on Demand GmbH, Copenhagen, Denmark
Printed: Books on Demand GmbH, Norderstedt, Germany

ISBN: 978-87-4300-241-3 Paperback,
ISBN: 978-87-4300-242-0 Paperback, black/white
Published as eBook

Table of Contents

1 Introduction

This book gives an introduction to the programming language **S**tructured **T**ext (ST) which is used in **P**rogrammable **L**ogic **C**ontrollers (PLC).

The book can be used for all types of PLC brands including Siemens Structured Control Language (SCL) and Programmable Automation Controllers (PAC).

The book is primarily composed to be used at the 2-year Full-Time Higher education AP Graduate in Automation Engineering and the Part-Time Higher education AP Degree in Automation and Operation.

In a Siemens PLC, the programming is called **S**tructured **C**ontrol **L**anguage (SCL) which includes some differences in relation to ST.

The book systematically describes the basic programming, including tips/advice and practical experiences from the author.

Many clarifying explanations to the PLC code and focus on the fact that the reader should learn how to write a stable, robust, readable, structured and clear code are also included in the book. Furthermore, the focus is that the reader will be able to write a PLC code, which does not require a specific PLC type and PLC code, which can be reused. It must also be underlined that the solutions can be used on the international market for automation solutions.

It is recommendable to read the entire book and then use the book as a reference.

Sorry, but no warranty on the PLC code examples in this book is provided.

1.1 Background for ST

ST is a high level programming language similar to Pascal Programming. Pascal Programming was widely distributed in Denmark from 1985 to approx. 2000 – a period of time in which many companies started developing software for PC, firstly DOS and since Windows.

ST is developed and published by International Electro technical Commission (IEC) in IEC 61131-3 International Standard in 1993. The standard consists of five PLC programming languages; where the LADDER Programming is the most well-known and most used.

ST programming for PLC Controls has since approx. 2010 been still more often published and since 2015 many companies in Denmark exclusively delivered PLC Controls, where ST is used as the favorite programming language. Consequently, still more employees are needed to understand and use ST, which is one of the arguments for distributing this book.

1.2 Qualifications for learning ST

It is not a necessary that the reader knows how to program in LADDER. However, certain knowledge of mathematics, mechanics, electronics, automation and basic PLC is necessary to be able to learn ST.

Students educated in a higher programming language (e.g. VB, NET, C, C#, Java) have the abilities to learn ST relative easy, as the programmable structures look like one another. The execution of the program in a PLC varies, however, a lot from a traditional PC program or a Web application.

The educational time for ST programming is like as other text programming languages expected to be from 3 to 5 years.

1.3   Foundation of knowledge

The author has 25 years of experience within specification, development and delivery of complex control systems and supervision systems. Of the 25 years, the author has 7-years of experience within Pascal Programing and 12 years within automation solutions and systems involving PLC. The experiences of comprises employment in four international companies and delivery of more than thousand control system solutions for 20 countries. Thus this experience provides an important basis for the substance in the book.

Within later years, the author has been teaching PLC Control Systems at higher educations. The students have from 0 to 20 years of practical or/and vocational experience within PLC, automation and technological service.

Furthermore, the internet, the standard DS/EN 61131-3 and series of books of PLC Control are applied as inspiration and clarification of ways to present problems.

The basis of the book is a material which is currently compiled with feedback from lecturers and students attending the AP Education in Automation Engineering at the local Dania Academy, 'Erhvervsakademi Dania', Randers, Denmark. The material is thus currently updated so that it answers all the questions which the students typically ask through-out the period of studying.

The author is Bachelor of Science in Electrical Engineering (B.Sc.E.E.)

1.4   Advantages of ST programming

ST is a very flexible and universal programming language. ST program code can easily be copied between different PLC types and be sent via e-mails as it is based on text and not graphics like the LADDER programming does.

The ST program code is similar to text sentences and work is carried out the same way as a word processor program (as e.g. Microsoft Word) which makes it easier to work on. Consequently, the same working methods are applied as in a word processor program.

Because of its very structured nature, ST is ideal for tasks based on complex math, code reuse or decision-making (e.g. automatic energy optimization, algorithms, data collection and regulation in process plants).

Having the experience with PLC Programming, the transition to other programming languages within PLC Control and automation will be easier; i.e. programming robotics or Visual Basic Programming.

Within later years still more companies have switched to ST Programming which is due to the fact that ST provides a series of advantages compared to the four other PLC programming languages (LAD, SFC, FBD and IL).

These advantages are as follows:

- ST Programming code can relatively easy be copied between different PLC types. [1]
- It is the easiest PLC language for mathematical calculations, formulas and algorithms [2] and large amounts of data (bigdata)
- PLC solutions are more demanding today than 20 years ago [3]
- Many widespread PC programming languages (C++, C#, VB, PASCAL, VB) reminds very much of the ST program structure.
- The other PLC languages (LAD, SFC, and FBD) require that parts of them are programmed in ST.
- It takes up less space when the PLC code must be documented, described and printed compared to the other PLC languages.
- It is the easiest PLC language to version control via comments in the program code or via GIT [4] or Subversion [4]

The PLC programming language Instruction List (IL) which is applied for complex PLC Controls is expected to be outdated within a few years (cf. DS/EN 61131-3 section 7.2.1) and it is expected that ST will replace these solutions.

[1] This is possible by using Copy-Paste and minor corrections. Siemens uses e.g. # before local variables and Allen Bradley another syntax for function 'calls'.

[2] Mathematical calculations are similar to mathematical formulas. See page 45

[3] Today there is more focus on energy optimization, automatic operation and data collection. These are all solutions which requires more complex PLC coding than merely an ordinary **'relay/circuit breaker'** with start/stop functions.

[4] The tools GIT and Subversion are practical tools in order to track (follow) corrections and extensions in the PLC Code. This makes sure that it is possible to fetch an earlier version (edition) of the right PLC Code in question.

1.5   Disadvantages of ST programming

A big disadvantage is the fact that many technicians and electricians are only capable of programming in LADDER. Further, it is difficult for them to understand the ST programming which is based on text and is not graphical as in the LADDER [1]

Programming in ST can easily be confusing as certain experience in structuring a program in an appropriate way is required.

Inexperienced people may have difficulties in fault-finding in a ST program.

Small (Micro) PLC does normally not allow ST Programming.

It is not possible to apply ST Programming in a safety PLC [2]

Reaching the expert level in ST programming often takes 3 to 5 years after ending the education/course.

2   How the PLC executes PLC code

It is important to know how a PLC executes a program which must be taken into consideration when the PLC program is written. A PLC executes programs sequentially in real-time, which means that the separate program parts must be executed within a short time. The program modules are executed at a fixed interval (PLC scan time) e.g. 50 [ms]. Some of the fastest PLCs may have a scan time of 1 [µs].

Program modules with different scan time e.g. 500 [ms] or each minute are possible. Sensors might occur which do not change their value quickly (e.g. a temperature sensor) and thus it is unnecessary to obtain quick scan time for all program parts. A large program including many calculations, takes longer time to execute and therefore it will be necessary to obtain different scan times for different program modules.

---

[1] In order to help people who are used to use LADDER programming to start ST programming instead, chapter 14, page 108, shows examples of chosen LADDER Programming and equivalent ST Programming.

[2] A separate PLC or special areas in an ordinary PLC are used to disconnect motors and other moveable parts if the emergency stop device is activated. The warranty must be a total of 100 % in order to have a proper disconnection and thus the PLC code is executed in a PLC in a safety mode, which is approved for this purpose.

The below flow diagram shows the basic mode of operation for a PLC:

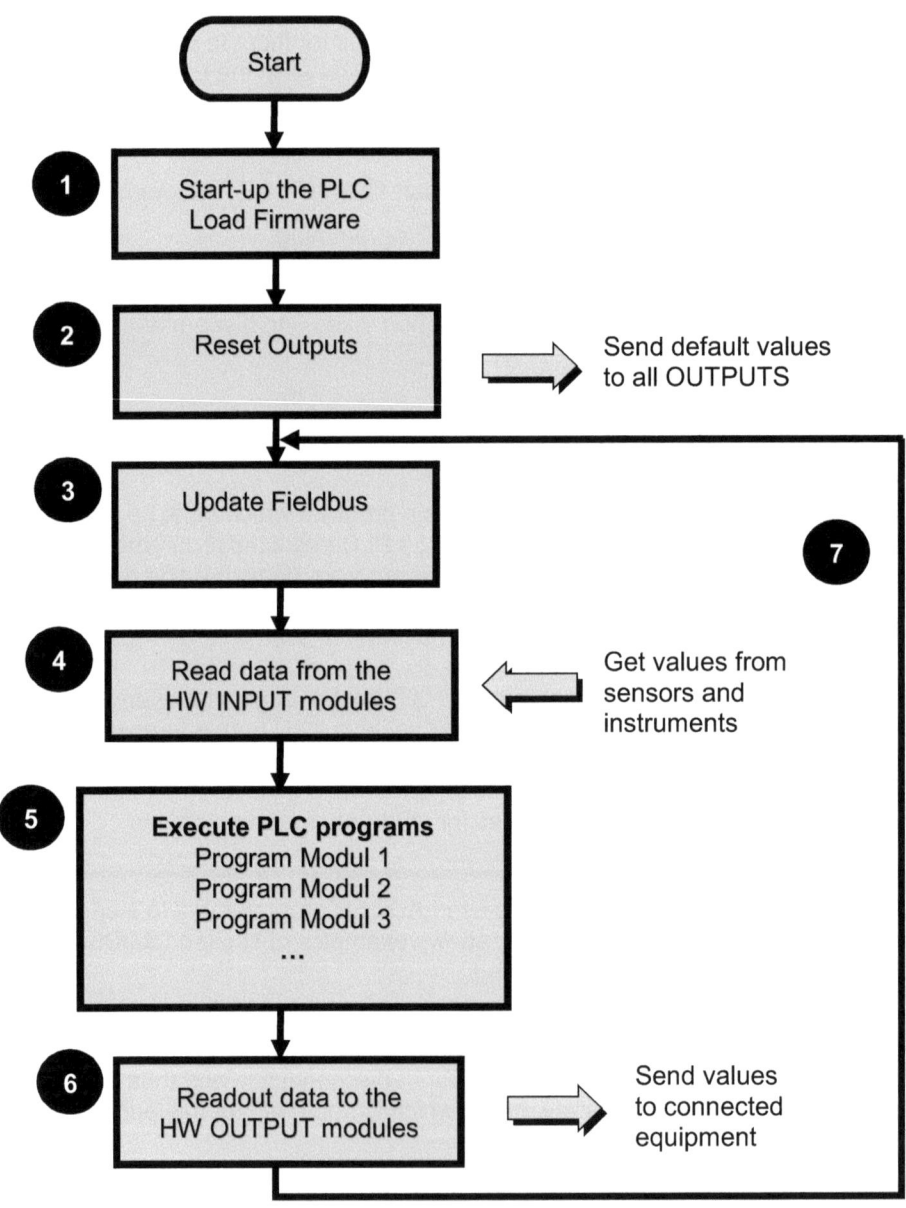

The flow diagram shows the following points:

1 When power is connected to the PLC it will start up / boot and load the operative system, named firmware in a PLC system. This will assure that the PLC program is familiar with the connected hardware (HW).

2 After startup all output modules are set to the value to which they are initialized. It is important that all outputs have the right startup value so that the machine does not carry out any unfortunate actions before the PLC program has started.

3 Now a data communication is made via a network (fieldbus). Hereby many variables are received and many are sent out to other units (e.g. control panels, other control systems or instruments). There are many types of Fieldbuses (e.g. Profibus, Profinet og EtherNet/IP). However, they basically have equal functions and working in the same way.

4 Now values from all sensors, contacts, breakers, instruments and components on the machine/unit are received from the input modules.

5 Execution of all PLC programs once, dependent on scan time. Programs are split up as follows:

> **Program modules**. Se section 10 page 68
> **Functions**. See section 10.1 page 69
> **Functions (FC) and Functions Blocks (FB)**. See page 72

Programs must be split up in order to create a good program structure.

6 Write values to all output modules, e.g. new settings to motors/engines, valves, lamps and instruments.

7 All sequences, points 3 to 6 will be repeated, which is one program scan.

The execution of the program only stops if:

- The PLC program is set to STOP mode
- If a run time error occurs
- The PLC is powed off or loses the power

3 Comments in programming code

Comments are a very important part of programming. Comments in the programming code assist you and your colleague when later adding to the code.

Use comments to explain what a specific PLC code performs, so you can remember it later yourself. In many cases, the PLC code can be self-explanatory, therefore it is a better choice only to make comments when coding is complex.

There are two types of comments in ST:

Line Comment

```
// Line comment. Forward-slash is written in front of EVERY line.

// Also used to remove/sort out PLC code – i.e. a code which is not executed
// The code has disappeared if it is deleted, therefore place // in front of the line
// instead of deleting the code. By doing this, the code is still to be seen
//but not executed
```

Block Comment

```
(* Block comment is initiated by start parentheses and a star. It is finalized by a star
and end parentheses. They are used for making more lines of PLC code inactive *)
```

Line comments can be only be positioned on the same line in front of or after code.

Comments positioned between (* and *) are named block comments and are used in order to remove/sort out more lines of code or to write comments filling up more lines.

For every program module or function, comments at the top lines are used so that another programmer quickly can read a description or introduction to the program module or the function.

It is recommended that the top lines contain a version log so it is possible to recognize what has currently been changed in the PLC code and by whom:

```
/////////////////////////////////////////////////////////////////////////////////////
/// OP002 Parking house
/////////////////////////////////////////////////////////////////////////////////////
// Action for each connected sensor
//
//************************************************
// Version 1.0, Created. Date 06.2.2018 TMA
// Version 1.1, TempVar3 changed 10.6.2018 TMA
// Version 1.2, Button B1 added 1.10.2018 TMA

IF B1 = TRUE THEN    //First line of PLC code
    K1:= TRUE;
END_IF;
```

A few PLC types cannot handle the special localized language letters such as æøå/ÆØÅ in the comment lines. It is therefore in general recommended to utilize the English alphabet in both the comment lines and the programming as localized special letters are often not accepted and, furthermore, many companies choose to write their PLC code in English. Another reason for writing the PLC code in English, is the fact that many companies are working internationally.

IMPORTANT   Do remember to correct the comments and version log
                 if anything is later is changed in the PLC code.

TIPS: A good idea could be to use comments to describe what the PLC
                 code could do, before starting to write the PLC code.
                 It could support the programmer to gain more structure and
                 again help the reader to understand the PLC code.

4 Data types

Just like other programming languages, the IEC 61131-3 standard provides many different data types, being both elementary and diverted ones. A data type defines how much memory capacity is needed by a variable value and by that, the largest and smallest value in the variable.

4.1 Elementary data types (INT, REAL, BOOL)

The following (chosen) elementary data types are standards in every PLC controller:

Data types	Bits	Numeral systems	Note	Value range Lowest and highest value	Example
BOOL (Bit)	1	Boolean (Binary)		FALSE/TRUE or 0 / 1	TRUE
BYTE	8	HEX (Hexadecimal)		16#0 to 16#FF	16#10
WORD	16	Binary		2#0 to 2#1111111111111111	2#0001000000000000
UNIT		HEX (Hexadecimal)		16#0 to 16#FFFF	16#1000
		BCD Binary-Coded Decimal		C#0 to C#999	C#998
		Integer without sign only positive numbers		0 to 65535	564
DWORD (Double word)	32	Binary		2#0 to 2#1111111111111111 1111111111111111	2#100000010001100 01011101101111111
		HEX (Hexadecimal)		16#00000000 to 16#FFFFFFFF	16#00A21234
		Integer without sign only positive numbers		0 to 4294967295 (4.29 billion)	435
INT (Integer)	16	Decimal Integer with sign		-32768 to 32767	101
DINT (Dou- ble integer)	32	Decimal Integer with sign		-2147483648 to 2147483647 (2.1 billion)	107

Data types	Bits	Numeral systems	Note	Value range Lowest and highest value	Example
REAL (Floating-point number)	32	IEEE Floating-point number (decimal value)	1	Lowest value: +/-3.402823E+38 Highest value: +/-1.175495E-38	1.234567e+13
LREAL (Long Real)	64	Double-precision floating-point IEEE 754		Lowest: -1.7976931348623E308 Highest: 1.79769313486232E308	3432.54
TIME (IEC time)	32	IEC time Step in 1 [ns] or Step in 1 [ms]	4	T#1ns to T#24d20h31m23s	T#10s T#10d14h11m23s T#5s12ms23us300ns
DATE (IEC date)	16	IEC day, step 1 dag		D#1990-1-1 to D#2168-12-31	D#2018-3-15 DATE#2018-3-15
TIME _OF_DAY (Time)	32	Time in a step of 1 [ms]	4	TOD#0:0:0.0 to TOD#23:59:59.999	TOD#1:10:3.3 TIME_OF_DAY#1:10:3.3
CHAR WCHAR	8 16	ASCII characters (letter)	2	'A', 'B' etc.	'E'
STRING		Text	3	Up to 255 characters	"This is a text"

All variables must have a data type. If a variable is given a value outside the minimum and maximum value range of the data type, a run time error may occur and consequently the PLC may stop the program execution. This may again lead to strange behavior when executing the program (the program may seem unstable).

A few PLC types provide more data types than the ones listed above. In general, it is recommended to stick to only a few data types so that the PLC code can be copied in an easier way to other PLC types. Some special data types such as S7TIME, LWORD and ULINT cannot be used by all PLC types. This indicates that copying PLC code having special datatypes or upgrading to a larger PLC may take a lot of work and a risk of introducing errors.

The three most commonly used data types are BOOL, INT and REAL. The reason why INT is more used often than WORD is that INT provides the same amount of data as the bit-size in a PLC and hereby it is a rapid data type. On the other hand, if REAL is used, the PLC will set up a machinery code behind the REAL, which is more complicated for the PLC to work on, as the PLC can only work with integers.

The disadvantage of working on INT is when INT is used in data communication be-tween two computers where one PLC applies a 16-bit and another applies a 64-bit operative system. It could even be a small 8-bit computer (an embedded computer), which could be a sensor, a measuring instrument, a device analyzing processes or yet another equipment on a facility. Read more about in chapter 8.5, page 49.

NOTER for the table

1)   A REAL integer contains max. 7 influential digits. Indicating that if a varia-ble is allocated to the value of 1234.56789 the variable is not able to con-tain all digits. The value will consequently be changed to the value of 1234.567 (7 digits). Some PLC types is using 8 digits: 1234.5678.

In some PLC types these data types are named FLOAT

As some computers interpret a REAL/FLOAT differently, some challenges can occur when communicating between more computers. In order to rec-tify this, a REAL 'is moved' to an INT or DINT variable by multiplying by 100 and when data are received in another computer the variable is divid-ed by 100. Hereby a decimal place including 2 digits can be transferred without any problems. See more in chapter 8.5, page 49.

2)   ASCII characters are typically used when texts are needed to be written on e.g. user interfaces, data logging to files, communication between instru-ments, data from a keyboard or other PLCs. Due to the fact that a PLC 'can only' operate with integers, letters and signs each have a number in an ASCII table.

The data type CHAR has 8 bits (may contain 255 different characters). A CHAR data type may typically be used for 1 to 5 different languages (countries). WCHAR has 16 bits and is applied for Unicode (ISO 10646, global signs). Unicode is meant for international PLC solutions.

WCHAR is typically used when the same PLC-code is applied in more countries with different languages in the user interface.

3)   A STRING consists of an ARRAY OF CHAR and is normally set to 255 signs (CHARS)

See also above-mentioned note 2). Furthermore, see chapter 4.5, page 23 and chapter 11, page 78. WSTRING is applied for Unicode (ISO 10646, global signs) and consists of an ARRAY of WCHAR

Note: Some PLC types provide a maximum of 80 characters in a STRING, if ARRAY is not limited to e.g. 10. It is good practice in programming to limit ARRAY so that not too much unnecessary memory is used.

When a variable is assigned to a value, the number normally (default) exists in the decimal system. If the value exists in the binary system **2#** must be stated in front of the number, and if it is a HEX number, then **16#** must be stated in front of the number. It could e.g. be **2#**101 = 5 or **16#**FF = 255.

Normally, an INT variable is used for counters and it is important to know when deciding how large a number can exist in an INT. If the INT is used as 'a time numerator' – TACHO HOURS on a motor (a counter showing the total amount of hours a motor has run and this specific number is used to see when service must be made on the motor). If the motor runs 20 hours a day and it has a life expectancy of 10 years, the total counter value will reach be as the following:

Hours within 24 hours*days per year*year = 20*365*10 = 73,000 (hours)

This causes the issue that the variable cannot be contained in the data type INT, as INT is   max 32767. A double integer DINT have to be used, or even better a DWORD datatype being able to contain an even larger number.

DWORD may contain an integer value from 0 to 4.29 billion.

If an INT is used anyway, the variable will show: 7466 as the INT has two 'overflows'. An 'overflow' takes place every time the integer is higher than 32767 and at an 'overflow', the variable is reset to -32768 (which is the lowest value for INT).

4.2   Introduction to derived data types

It is possible to define more advanced and adapted data types to save time when programming and to obtain a better program structure. The data types are named derived data types and are declared within **TYPE** and **END_TYPE**.

The four datatypes are as follows:

- Structured data type, **STRUCT,** See chapter 4.3, page 19
- Numbering data type, **ENUM,** See chapter 4.4, page 21
- Sub-Range data type, See chapter 4.5, page 22
- Range of identical data type, **ARRAY,** See chapter 4.6, page 23

NOTICE

If an absolute beginner starts programming in a PLC, it is important to know that the first three data types are not necessary to use, in order to make PLC programs work. This means, start using derived data types when greater experience in PLC programming is gained.

The different derived data types are explained in the following chapters.

4.3   Structured data type, STRUCT

A structured data type STRUCT is a composite data type used for collecting more datatypes in a group (Class/object). The structured data type is created by using the key words TYPE, STRUCT and END_STRUCT.
Each variable in a STRUCT needs to have an indicative name followed by a colon and then the data type. Notice that the expression is ended by a semicolon.

Below here a STRUCT is shown, called **Motor**, containing four variables which are all related to a motor. **Speed** (on the motor) **Temperature** (measurement in the motor), **Voltage** (Power supply) and **AlarmStatus:**

```
TYPE Motor :                                  //Example 1 STRUCT
        STRUCT
        Speed         : INT;    //Actual speed of the motor [RPM]
        Temperature : REAL;  //Internal temperature of the motor [C]
        Voltage       : REAL;  //The voltage of the motor [V]
        AlarmStatus  : BOOL;  //Alarm if TRUE else FALSE
        END_STRUCT
END_TYPE
```

Notice that comments are written after each variable which precisely describes what it means thus the reader of the PLC program is not in doubt. Furthermore, a unit is quoted in square brackets because the unit of different variables can often not be known, e.g. the speed of the motor is measured in RPM (revolutions per minute), the frequency in Hz (Hertz) or in percentages (0 to 100%).

Comment lines, when the variable is set up, are also used for describing what the value of the variable can be, as it is not always logical; e.g. at **AlarmStatus** when it is not clear whether it is an alarm when the variable is **TRUE** or **FALSE**

As mentioned in chapter 6.1, page 34, the unit can be a part of the variable name.

Some PLC types do not use a text, as mentioned above, to set up a STRUCT; they are, however, set up on a list and therefore TYPE, STRUCT, END_STRUCT or END_TYPE will not be seen.

An actual structured data type may contain one or more structured data types. This can be seen in the example below:

```
TYPE Valve :                                 //Example 2 STRUCT
    STRUCT
    DisplayColor   : LightTYPE; //User defined TYPE
    ValveState     : BOOL;        //Can be TRUE (open) or FALSE (closed)
    Pressure       : REAL;        //Pressure in [Bar]
    END_STRUCT
END_TYPE
```

In example 2 above the data type **Valve** consists of tree variables:

DisplayColor, **ValveState** (Status for the valve: open or closed) and **Pressure**. The variables **Pressure** and **ValveState** uses the standard data types **REAL** and **BOOL**, while the variable **DisplayColor** uses the data type **LightTYPE**, which is defined in the following chapter 4.4, page 21.

An example of a portable tank containing chemicals (IBC tank):

```
TYPE TankType :                 //Example 3 STRUCT
    STRUCT
    Liters  : REAL := 1000;  //Default tank size
    LevelSensor   : REAL;     //Sensor at bottom
    LevelSwitch   : BOOL;     //Float switch at bottom
    END_STRUCT
END_TYPE
```

Many variables in a PLC program may easily seem confusing. Variables belonging to an affiliation of the same component (object), the same domain, or the same mode of operation, may advantageously be collected in a STRUCT. It is consequently easier and quicker to set up and maintain many identical components. This method is called Object Oriented Programming (OOP) and is typically recognized from programming a PC, the method can profitably be used in a PLC.

If a variable having the data type STRUCT is to be transferred to a function the variable scope must be VAR_IN_OUT in the function. See chapter 5, page 26.

4.4   Numbering data type, ENUM

The numbering data type **ENUM** contains a list of unique names. Names are listed in a parentheses. Expressions begin with TYPE and closes with END_TYPE. Indicative names must be used thus they explain for which purpose they are used.

As an example:

```
TYPE LightTYPE :
      (RED, YELLOW, GREEN);
END_TYPE
```

The data type **LightTYPE** in the above-mentioned example can either be indicated with a RED, YELLOW or GREEN sign, or can e.g. be applied for a traffic light, an operator signal lamp (see the picture) on a machine or as a status on a valve. **LightTYPE** will/must always be indicated by one of the defined types: RED, YELLOW or GREEN.

An **ENUM** *must be allocated as* a default; otherwise, it is uncertain what the initializing (commissioning) value is. As an example, shown below, **LightTYPE** will have the value RED, when the PLC is powered up:

```
TYPE LightTYPE :
      (RED, YELLOW, GREEN):= RED;
END_TYPE
```

The PLC-compiler (Program which converts the ST program code to the PLC) automatically inserts a consecutive number for each individual text: RED = 0, YELLOW = 1 and GREEN = 2, based on the fact that a CPU can only work in numbers. Hereby the name **ENUM** occurs; **ENUM** (enumeration) can be translated to 'automatic numerical order'. The method is used because it is easier for the programmer to remember a text instead of a constant number and the programmer needs not to use time on writing the underlying consecutive numbers.
It is possible to define a fixed value at each name instead of using the consecutive ones as shown:

```
TYPE LightTYPE :
      (RED:= 10, YELLOW:= 20, GREEN:= 30) := RED;
END_TYPE
```

The disadvantage of using **ENUM** is that all numbers are positioned in a continuous order. If new names are added in the middle of the sequence, the order of numbers is displaced and consequently creating new challenges when **ENUM** variables are exchanged between more PLCs or computers, as both must be updated with a new PLC code at the same time.

Examples of use: Below are two variables, **MotorLamp** and **Lamp**, both having the data type **LightTYPE**:

```
Lamp:= MotorLamp;                //Here Lamp is set to red
MotorLamp:= LightTYPE.green;     //Set MotorLamp to green
Lamp:= MotorLamp;                //Here Lamp is set to green
```

ENUM creates a better structural software, but **ENUM** is not possible in all PLC types.

The alternative to **ENUM** is to use independent constants. See chapter 6.2, page 36.

4.5 Subrange data type

A sub-range data type is a data type, delimited in relation to an elementary data type. This makes sense in case of a restricted measuring area. A subrange consists of the name of the delimited data type, followed by a lower and an upper limit, divided by two points or dots, both of them noted in a bracket, which is shown below as an example. Here we see the delimited data type **TemperatureRangeType,** which is only able to contain numbers between -50 and 125.
(The INT data type is able to contain values in the range of -32768 to 32767):

```
TYPE TemperatureRangeType:
        INT (-50 .. +125);
END_TYPE
```

If a variable with the data type **TemperatureRangeType** is allocated a value 132, consequently outside the permitted range, a run time error can occur in the PLC. Therefore, Subrange data types are not often used because run time error is not very simple to handle (and to explain to the customer) compared to a variable showing 132 instead of 125. If the value is visible, it can be simpler to notice that the value lies outside the permitted range and then it is easier to locate the problem.

4.6   Range of identical, ARRAY

An ARRAY is a structured method utilized for saving more values with *the same data type.* The positions are situated side by side in the memory indicating that it is simple to work on. An ARRAY always provides a pre-determined fixed length which cannot be changed during the execution of the program. An ARRAY can be set up and indexed in several dimensions. A PLC code is easily connected to an ARRAY and indicates a profound software structure. The challenge is based on getting the values in and out of ARRAY.

An ARRAY is also called a multi elementary data type.

Below an example is shown, where **SpeedArray** contains 6 positions of the data type INT. To define the 6 positions use ARRAY followed by square brackets including start position number and end position number, divided by two dots as shown below:

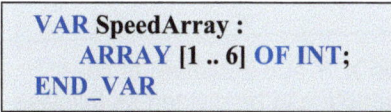
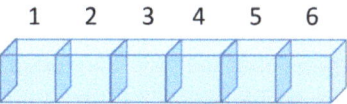

The first value in the array is located in position no. 1 and the last value in position no. 6. A naming is now chosen where **Speed** is added to the text **Array** so that the person working on the PLC code may easily notice that it is an ARRAY which is now in use.

SpeedArray is a one dimensional ARRAY and can be applied where a collection of many values is positioned in one long row, as follows:

Calculation of the average value (chapter 10.4, page 66).
Handling of a queue (chapter 13.1, page 92).
FIFO - **F**irst **I**n **F**irst **O**ut (chapter 13.2, page 95).
Collection of data and sorting (not included in this book)

An ARRAY can be set up with all data types, including STRING, STRUCT or functions.

An example of the use of ARRAY can be seen on page 64, 66 or 92.

A two dimensional ARRAY can be used on e.g. a parking lot, stock rack, a graph, a bar chart or a pivot table and can be set up as follows:

```
VAR Racking
    ARRAY [1 .. 5, 1 .. 3] OF INT;
END_VAR
```

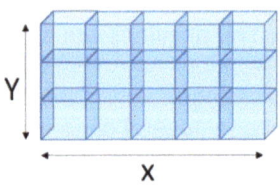

A three dimensional ARRAY is defined as follows:

```
VAR PackOnPallet
    ARRAY [1 .. 5, 1 .. 4, 1 .. 3] OF REAL;
END_VAR
```

Used e.g. for packages on a pallet (palletiz-ing) or positions in a warehouse

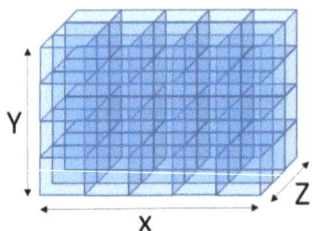

If a three dimensional ARRAY is in focus, such as an X, Y and Z system of co-ordinates, the values from the above-mentioned example can be divided as follows:
    1 to 5 = X; 1 to 4 = Y; 1 to 3 = Z.

The total amount of positions in **PackOnPallet** ARRAY is: 5*4*3 = 60 pieces. Thus, this ARRAY contains 60 positions.

An ARRAY can be defined to start from 0. The ARRAY below contains 4 positions, as the position 0 (zero) and position 3 are included. It results in a more stable program when arrays is starting from 0, because the array index pointer can be uninitialized.

```
VAR MyArray1D
    ARRAY [0 .. 3] OF INT;
END_VAR
```

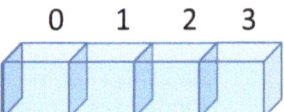

Insert a single value in an ARRAY

When values in a one dimensional ARRAY are allocated, it is carried out as follows: Below, the value 5 is inserted at position 4 in the array named **SpeedArray**.

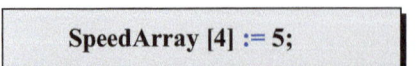

```
SpeedArray [4] := 5;
```

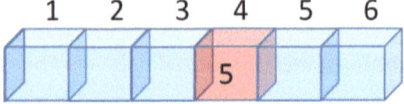

© 2019 Tom Mejer Antonsen

Insert values in a 3 dimensional ARRAY
PackOnPallet as follows:

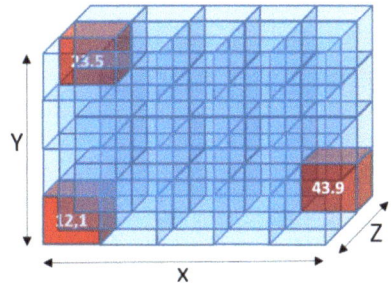

```
PackOnPallet [1, 1, 1] := 12.1;
PackOnPallet [5, 1, 3] := 43.9;
PackOnPallet [1, 4, 2] := 23.5;
```

If inserting many values in a 3D ARRAY, see chapter 9.4.2 page 65

Take-out values from an array

Now a value is collected from a one dimensional ARRAY. The value is located at
position 2 in the array named **MyArray1D** and copied to the variable **Var1**.

```
Var1 := MyArray1D [2];
//Value in Var1 is 12
```

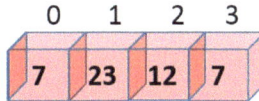

Taken out a value from a 3 dimensional ARRAY is carried out as follows:
A value is transferred to the variable **Var3** having the value of 43.9:

```
Var3 := PackOnpallet [5, 1, 3];
//Value in Var3 is 43.9
```

IMPORTANT: Areas outside the ARRAY domain must not be written to. If a writing to
e.g. a position no. 10 is attempted in an ARRAY containing only 6 positions, the PLC
can stop the program execution (Run Time Error). This is a typical error/mistake
which is often made.

The way to avoid this is to secure that changes/operations in the ARRAY are only
carried out, when an IF condition is fulfilled as shown below:

```
Index:= 4;
IF Index > 0 AND Index <= 6 THEN
    SpeedArray [Index] := 5;
END_IF;
```

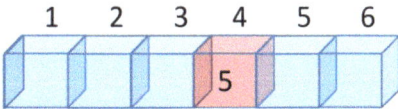

5 Variable scope

Variables are key elements in the programming. All variables each have a data type.

When a variable is created, it must be connected to a variable scope describing the value's behavior in the memory.
A table of the most typical variable scopes in the PLC program is shown below:

Scope	Description
VAR	Between the keywords VAR and END_VAR, all of the local variables are declared. The local variables cannot be manipulated from outside the program module or the function. TIPS: In some PLC types VAR is replaced by *Static*
VAR_GLOBAL	Global variable scope. Variables which this scope can be used from all program modules, functions, Fieldbus (network communication) and HMI (Operator user interface) Limit the use of global variables, because it makes the PLC code more complex and it will be difficult to find errors
VAR_INPUT	Used by functions for variables *entering* a function See more in chapter 10.2 page 72
VAR_OUTPUT	Used by functions for variables *exiting* a function. See more in chapter 10.2 page 72
VAR_IN_OUT	Input and output variable scope for functions. An address of the variable *transferred* to the function and work is done directly on the variable and *not as a copy* as when working on VAR_INPUT Used when a function have to work with a STRUCT or ARRAY The scope must be used carefully as the function changes in a variable located outside the function. See more in chapter 10.2 page 72
VAR_EXTERNAL	If a program module use this scope on a variable, the program module will be able to use the global variable of the same name. Must be used with caution
VAR_TEMP	A temporary variable scope in the function indicating that the contents of the variable disappears when the function is ended

Scope	Description
AT	Allocate a memory position (address) for a variable. It can be an IO address (the address on a PLC input or output). The input can be named %IX 1.0, where %I explains that it is an input or %QX 0.0, where %Q explains that it is an output. Q is used as a letter for output (O is not used as it can be mixed up with zero/nil). See the example in chapter 5.1 on page 28. If nothing is indicated, the PLC will normal automatically allocate the next free internal address in the memory.
CONSTANT	This variable cannot be changed during runtime. Used for numbers and values which must be fixed through the whole program. It is important to use this variable scope, when the same fixed value is used *more than once in the same PLC code*. See more in chapter 6.2 on page 36
RETAIN	Retain the variable value after a possible power failure or power loss. The variable is saved in memory (the internal memory). It is IMPORTANT to use a variable containing e.g. hour counters, subject counters or similar ones. As these values must not lose the value they have reached if the PLC is turned off. Cannot be used in a FUNCTION
PERSISTENT	As RETAIN. But the variable is saved in a file on the hard disc. It is IMPORTANT to use this for values containing e.g. hour counters, subject counters or similar ones. This is often only possible to use in a soft PLC. Further, this makes it simple to move the contents of variables to another PLC, e.g. if a PLC has to be exchanged. Cannot be used in a FUNCTION
END_VAR	End of the variable scope statement Default (required)

5.1 EXAMPLE: Variables, Scope and IO-modules

This chapter shows an example with a variable creation:

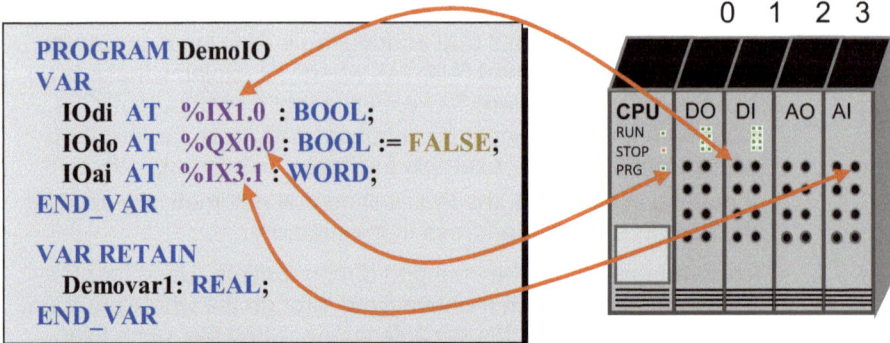

The example above shows four local variables in a program module, named **DemoIO.**

There is a variable **IOdi** with the data type BOOL having direct connection to port address no. 0 on the hardware input card no. 1. It does not make sense to initialize the variable, as the value is determined by the sensor, which is connected to card.

The output variable **IOdo** is default set to **FALSE** to be sure that the output signal is set to zero when the PLC is power on. It has a direct connection to port address no 0 on hardware output module) (the card closest to CPU)

The input variable **IOai** is an analogue value with the data type WORD. An analogue input value can be 16 bit, but typically 12 or 13 bit, as they are cheaper and a 16 bit resolution is not always necessary. The variable has a direct connection to port address no.1 on hardware output card no. 3

Furthermore, **DemoIO** has a local variable **Demovar1** with the data type REAL. **Demovar1** is saved in case of power failure or power turned off, as it is marked with RETAIN thus a counter value is retained.

Some PLC-types do not have a direct address on input or output card as shown above. In the PLC types, %I* and %Q* are written, as well as in a mapping table, a list of connection between variables and the physical input and output card, where it is possible to connect variables with the physical input and output card.

6 Naming of variables

Naming the variables (tags) is important. This chapter and the following, go through rules and methods for naming the variables reasonably.

Each company often has its own rules and attitudes of how naming must be, which is the reason why some guidelines and examples are mentioned below. Many PLC programmers do also have their attitudes of how precise naming must be. The most important is an indicative variable name followed by a comment in the section where the variable is created.

Variable names must begin with a letter, where after the name can contain combinations of letters, numbers and some symbols, such as *_*.
Variable names must not have the same names as pre-defined functions, standard routines or user-defined functions. Variable names such as **ARRAY**, **REAL** or **INT** are therefore invalid.

PLC system requirements for variable names:

- Invalid signs: ~ @ ; " # % & * : < > ? / \{ | },. SPACE, TAB
- Often invalid local language letter like Danish special signs: æøåÆØÅ
- Use short indicative names: Some PLCs have max. amount of 24 letters
- Variable names must not start with a number
- Take care not to use the letter O close to a number
- There is no difference in using lower and CAPITAL letters (lower/upper case)

TIPS when naming with more words: First noun and then verb.

E.g. **PumpRun**, where **Pump** is a noun and **Run** is a verb.
If a word has two nouns, begin naming with the big component:
E.g **PumpSensorError** or **TankSensorLevel**

There are four methods within naming the variables:

Hungarian Notation
Camel Case
Pascal Case
Snake Case

Hungarian Notation

This notation means that the letters: i, s, ar, b, are inserted in front of the variable name in order to tell the programmer which data type is used. However, some unfavorable circumstances can occur, if the variable later shifts to another data type, as variable names then must be changed in both the PLC code and the belonging documentation. Furthermore, many programming tools today show the data type of a variable with a Tool tip function (a small yellow box, appears if the computer mouse is held over the variable name).

Examples of used letters: x = BOOL, i = INT, I = REAL, ar = ARRAY, s = STRING, b = Bit, w= WORD, jw = DWORD, e= ENUM

Examples:

iMotorSpeed	(Speed on a motor with the data type **INT**)
xMotorAlarm	(Alarm on the motor defined as the **BOOL** data type)
sMotorAlarm	(**STRING** containing a motor alarm text)
arMotors	(**ARRAY** with motors)

Camel Case

This naming rule uses notions from Camel Case where the variable name is beginning with a lower letter and the following words are written with capital letter.

Examples:

flowMeasureWarningBit	blowerStartBit
motorSpeed	calculateError
sensorHighSignal	motorInitFunction
sensorLow	powerEstimated

Pascal Case
This naming uses notions from Camel Case where the compound of names is always starting with a capital letter.

Examples:

FlowMeasureWarningBit	BlowerStartBit
MotorSpeed	CalculateError
SensorHighSignal	MotorInitFunction
SensorLow	PowerEstimated

This is probably the most often used method today, as it is easy to read, quickly to write and creates the shortest word.

Snake case

In this method, underscore signs are used to differentiate words. Underscore signs are used as the <SPACE> sign is not valid for naming in the PLC. It can be difficult to read, when underscore is used and creates too long a name. Some PLC-types allow a maximum of 24 signs in a variable name and can creates a challenge when variable names become very long:

Examples:

flow_measure_warning_bit   blower_start_bit
timer_done_bit   calculate_error
initial_motor_frequency   motor_init_function
sensor_high_signal   power_estimated

A big advantage in using Snake Case is when tools for automatic generating of TAGS/variables are used in IO-Lists, electrical diagram drawings, PLC and SCADA codes, as "_" can easily be exchanged with "." via search-and-replace routines.

If an abbreviation is made, <u>use</u> only standard abbreviations, such as **Cal** for calculate **avg** for average or **Cmd** for command.

If specific company abbreviations or own abbreviations are used, a comment must be written in the code, or where the variable is created, because otherwise it might be difficult for other readers to find out what the abbreviation means.

Below you see two identical PLC codes examples where Pascal Code and Snake Case are used respectively for creating variable names. Consider which PLC code is the simplest to read (the one to the left or to the right):

```
IF TankLevel >= EmptyLevel THEN          IF tank_level >= empty_level THEN
    ValveOpen := TRUE;                       valve_open := TRUE;
    IF ValveError = TRUE THEN                IF valve_error = TRUE THEN
        ValveOpen := FALSE;                      valve_open := FALSE;
    END_IF;                                  END_IF;
END_IF;                                  END_IF;
```

Which one to use?

Which of the naming methods is the best one is often a matter of attitude and can be decided by which method you earlier used.

It is also important to choose meaningful words for variables. An example could be a variable which must show a status for pump no. 141. This could create the names as follows below:

> Pump_Status_141, Status_141, P141_Status, Pump141Status, PumpStatus_141, P141S, ...

Pump141Status is the best choice as the noun is the first name mentioned. The number of the pump (141), suited for the noun, appears after the noun (Pump) and finally the verb (Status). Furthermore, Pascal Case is chosen as naming method, because it creates short readable variables names.

Variables including only one letter i, j, x, y, z, k, n as iterative variable (e.g. counters and loops) and index/pointers in **ARRAY** are often used. It is easier to write a single letter than to write e.g. **ArrayIndex**. Often x, y, z are used in systems of coordinates.

Variable names such as **Temp1** and **Temp2** can be applied as contemporary variables, which should not be used very often, because they are not very indicative.

Variables with names containing words such as **New** and **Changed** must be used carefully, because they are not new to the programmer who is correcting the PLC code later on.

Some programmers prefer to use the data type, as a part of the variable, e.g. Int_Number_of_Run and Real_Initial_Temperature. It reminds a little of Hungarian Notation and creates some long names and can cause some trouble, if the data type is to be changed later on.

A further advantage could be adding a unique number in front of each variable name, thus the variables are simpler to identify and look for in the code and documentation:

B8040_MotorSpeed	S213_PumpAlarm
B8041_MotorCurrent	S101_PumpSpeed
B8044_MotorPower	S001_SoftwareVersion

The above-mentioned methods are naming of variables. The methods can, however, easily be used for naming functions, function blocks, and program modules.

Some programmers use **fb** and **fc** in front of their own functions and function blocks:

fbCalculateArea	**fcArrayFindMin**
fcMotorStatus	**fcArrayFindMax**

Many built-in standard functions and routines do not use **fb** and **fc** in naming, which makes it difficult to be consistent.

Naming such as B1, B2, B3, B4 …. is not OK, unless the naming is used in the problem statement (The control requirement specification or functional description)

Naming of **STRUCT** (see chapter 4.3, page 19) can advantageously be added *TYPE* to the name, thus it is easy to see that **STRUCT** is used.

Alarm texts can both be of the data type **STRING** consisting of a text and **INT** if an alarm number is in question. Many user interface languages can be included on the control panel, thus naming can be as follows:

sAlarmMotorLoad_DK	"Alarm motor overbelastet"
sAlarmMotorLoad_UK	"Alarm motor overload"
iAlarmMotorLoad	12004

When naming, many companies within the process industry (dairies, breweries, the medical industry, the oil industry) follow the S88 standard (ANSI/ISA-88). It indicates a naming depending on the sensor type and the installation location. Naming spans over the IO list, the control specification, the PLC program and test documents, which makes it easier to overview all variables, the PLC code and the documentation. Using the same naming of the variables in the entire control system obviously creates fewer misunderstandings and a better quality.

Examples:

FZ.MM01.UE01.PO3	FZ_MM01_UE01_PO3
FZ.MM02.UE01.M01	FZ_MM02_UE01_M01
FZ.MM02.UE01.TT01	FZ_MM02_UE01_TT01

Where **PO3** is "Control Module", **UE01** is "Equipment Module" and **MM01 is** "Process Celle" according to naming rules following the S88 standard.

6.1   Variables with unit

Many variables must have a connection to a unit; otherwise it does not make sense. If a variable e.g. is created to represent a temperature, the temperature must be stated in °C (degrees Centigrade/Celsius) or °F (degrees Fahrenheit (USA)). In order to facilitate the programmer's job, the right unit can be informed by adding it to the variable. A variable measuring temperature in degrees Centigrade can e.g. be noted as **MeasureTemperatureC**, where °C indicates the unit, which can even be noted in the commentary field where the variable is created.

Examples of other (chosen) variables needing a unit:

Variable	Chosen possible units
Time, period #1)	us, s, seconds, minutes, hours, days, week, year
Speed	m/s, km/t, km/h, rpm, %, mph, mm/s, tf/s
Amount	kg, g, no., kr., dollars, pcs., liters, bottle, box
Weight	kg, pounds, lbs., g, tons, mg, %
Oxygen	mg/l, %, g, l
Consumption	W, kWh, kr, l, kg, $, m, m2, m3, A, k/j, g, l/h

It is important to obtain an overview of all units and in some PLC Control Systems it is required that the PLC Control System itself is able to change units when the consumer wants to, especially if the same PLC Control system is applied to be used globally. E.g. a temperature indication in °C or °F . It is typically the fact in USA and Canada where you are required to be able to change the unit on temperature values online.

The conversion between °C and °F is a formula to be found on the internet. Here it is shown how to calculate °F based on °C temperature:

VarF:= (VarC * 9/5) + 32;

Units can be SI-units; notice the SI-prefix.

The view on user interfaces (HMI) and data log for files and reports, etc. must often include max. two decimals. If a value is to be shown with two decimals on an HMI, this is often done by written **%f5.2** in the text field on HMI. The **%f** means a FLOAT (**REAL**) value and **2** means 2 digit after dot as shown:

$$23.45 \; [°F]$$

Units are often written with square brackets to increase the readability; e.g. temperature [°C]. This is the fact both in HMI and its belonging documentation.

It is common to create a function for the conversion between different temperature scales, because it is the same specific PLC code which must be used several times and for other customers. An example of this is shown at page 75.

NOTE: Time, periode #1)

Time can be a difficult factor to work on in a PLC used internationally. There is a difference in which day of the month the countries shift between summer and winter time and whether Sundays are the first or last day in a week. Finally, there is a difference in when week no. 1 in calendar year starts.

Examples of variables having unit as a part of the name:

```
TemperatureC
TemperatureF
MotorSpeedHz
MotorSpeedPercent
ConsumptionW
ConsumptionKWH
MotorUseA
```

6.2   Variables with fixed value (CONSTANT)

Variables, fixed and unchangeable during the programming process, are indicated as a constant value (**CONSTANT**). They are used for the same numbers, *used more than once in the same PLC code*. This ensure that the same used numbers are corrected everywhere.

• The **CONSTANT** variable names are often written in CAPITAL letters (Upper-Case).

When must a CONSTANT be created?

If the PLC code must be multiplied and divided several times, e.g. by 25.4, which is the converting factor between millimeters and inches, a constant must be defined; e.g. **MILLIMETERS_PER_INCH** = 25.4 and it is used everywhere in the PLC Code. On the other hand, it is not likely that the converting factor between millimeters and inches must be changed. If it is to be changed, 25.4 can easily be changed with a 'search and replace' function. Furthermore, it takes longer time to write a long text than to write 25.4. Possibly, other 25.4 values can be changed when the 'search and replace' function is carried without intention. The consequences are very unfortunate. If constants are used with names, the consequence is a self-explanatory program.

The definition of the length in connection with the creation of **ARRAY** must always be defined as constants, because they are used several places in the PLC code. It Create an unstable program, if not all definitions are changed when the length of **ARRAY** is changed. The length of an **ARRAY** is changed when e.g. testing an **ARRAY**. As an example see page 66, where **BufArrayMin** and **BufArrayMax** are created as constants and they are used together with an **ARRAY** named **BufArray**. By adding *Min* and *Max* to the Array name, it is clear to see that they all belong together.

Arguments for using constants:

1)   The PLC Code is more readable
2)   Avoid errors when changing constants and numbers
3)   Save time when changing a number

Examples of using constants:
PI:=   3.1415927
SECONDS_DAY:=   86400

7 MATH and LOGIC

The following chapter deals with operators found in math and logic together with the mathematical functions built into a PLC Control.

7.1 Arithmetic Operators (+, -, *, /)

Table of ordinary arithmetic operators (mathematical symbols):

Operator	Explanation	Functions	Examples where V1 = 2 V2 = 5	? Y =
+	Addition	Y:= **ADD**(V1,V2);	Y:= V1 + V2;	7
-	Subtract	Y:= **SUB**(V1,V2);	Y:= V1 – V2;	-3
*	Multiply	Y:= **MUL**(V1,V2);	Y:= V1 * V2;	10
**	Exponent	Y:= **EXPT**(V1,V2);	Y:= V1 ** V2;	32
/	Divide	Y:= **DIV**(V1,V2);	Y:= V1 / V2;	0.4
MOD	Modulo	Y:= **MOD**(V1,V2);	Y:= V2 **MOD** V1;	1

Where V1, V2, Y can be variables or numbers (integer or decimal numbers)

The built-in functions **ADD**, **SUB**, **MUL**, **EXPT** and **DIV** can easy be used. But it does, however make, better sense in the ST-programming to use the sign mentioned in the **Operator** field in the table above, as it is similar to 'ordinary' mathematics and the method used in math formulas and other calculation programs.

Not all PLC types support ** operator. Use the **EXPT** function: Y:= **EXPT** (V1, V2);

One of the forces in ST programming is that the math calculations are similar to the methods used in math-programs and consequently the calculations are simple to write, to troubleshoot on and read directly in the PLC code.

Examples of using the math operations can be seen on page 43 and page 98.

In order to perform calculations, it is important to choose the right data types for the variables. In most cases, a **REAL** variable will be a suitable data type.
If **INT**, as an example, is used as a data type, the calculation can in some cases create variable overflow, because of a data type, which is too small or is wrong. This is due to the fact that calculation results in a large number.
This can be illustrated as in the following example:

Calculating: **Y = V1**V2**, ($Y = V1^{V2}$)

where V1 = 10 and V2 = 10, results

Y = 10000000000

The value Y cannot be an **INT** data type.

IMPORTANT

Choose a suitable data type for the calculation.

If too large a data type (e.g. **LREAL** or **LWORD**) is chosen, more memory is used and it has a more impact on the PLC Controller than necessary.

© 2019 Tom Mejer Antonsen

7.2   Relational Operators (=, <, <=, >, >=, <>)

In order to compare the relation between two values (integer or decimal numbers) it is possible to apply relational operators. The two values can be variables or numbers. The result of the comparison is a value, which always has the data type Boolean (BOOL) and can therefore only be TRUE or FALSE.

The relational operators are as follows:

Operator	Description
=	Equal, same
<	Less than
<=	Less or equal
>	Greater than
>=	Greater than or equal
<>	Not equal, not same

Example of use:

```
HeaterOn := Temperature < SetPoint;
```

The data types for **Temperature** and **SetPoint** are both REAL. The expression can be used if e.g. a heat lamp has to be switched on, if the temperature is too low. **Temperature** can be measured by a sensor connected to an analogue input module. The temperature at which the heat lamp has to be switched on, is set on **SetPoint.**

Explanation of the above example:

HeaterOn will be TRUE, if **Temperature** is lower than **SetPoint.** As the expression **Temperature < SetPoint** result in a variable of the data type BOOL, **HeaterOn** must be a BOOL data type. The variable **HeaterOn** can be connected to a digital output module, which when TRUE activates a relay that turns on the connected heat lamp.

Relational operators are mostly used in connection with IF-statements (PLC programming lines having IF included), see chapter 9.1, page 53.

7.3 Numeric Operators (MATH functions)

This chapter describes the built-in math functions in a PLC.

Math functions have typically only one input parameter being a number, typically a data type, INT or REAL. A return parameter from the function often has to be of the data type REAL. It is important to secure that the input parameter is valid. It is e.g. not possible to call the LN function with the value 0, as it is not mathematically correct and the PLC Controller stops the program execution (Run Time Error).
A correct program execution can e.g. be carried out as follows, where **x** is an input parameter and **y** is the result when calling the LN function:

```
IF x <> 0 THEN
    y:= LN(x);  //Only executed if x is not zero
END_IF;
```

A table of the built-in math functions in a PLC is seen below:

Function	Mode of operation (Example where a = 2, b = 5, c = 8)
NEG	Change a positive number to a negative number and vice versa. Same as a:= a * -1;
INC	Count 1 up. Add 1 to the value. Increment, INC(a) = 3. The same as a:= a + 1;
DEC	Count 1 down, Decrement. DEC(a) = 1. The same as a:= a - 1;
TRUNC	Converting a REAL value to an INT value. No rounding off the integer value. TRUNC(3.9) = 3 TRUNC(-2.5) = -2 Value after dot is removed
FRAC	The decimal value of a value. FRAC(2.8) = 0.8, FRAC(-3.49) = -0.49
ABS	Absolut value. The function insure always a positive value. ABS(-1.2) = 1.2 ABS(3.4) = 3.4 ABS(-3) = 3
FLOOR	For positive values, the returned value is less than or equal to the input For negative values, the returned value is greater than or equal to the input. FLOOR(2.8) = 2 FLOOR(-2.8) = -3

SQR	Square. This function calculates x^2, raising to the power of 2. SQR(4) = 16 The same as x * x , (x multiplying x)
SQRT	This function calculates the square root. SQRT(4) = 2, SQRT(9) = 3
LN	The natural logarithm. LN(2.71828) $\approx$ 1 (the wave sign means approx.)
LOG	The natural logarithm with base 10. LOG(10) = 1.
EXP	Exponential function. Same as e^x or e^x, e = 2.718281828
SIN	Sinus function. SIN(a) = 0.35 (GRAD) **#1)**
COS	Cosines function. COS(a) = 0.99939 (GRAD) **#1)**
TAN	Tangent function. TAN(a) = 0.03492 (GRAD) **#1)**
ASIN	Arc sinus function. Inverse sinus function. $SIN^{-1}(x)$, Sinh(x). **#1)**
ACOS	Arc cosines function. Inverse cosines function $COS^{-1}(x)$, cosh(x). **#1)**
ATAN	Arc tangent function. Inverse tangent function $TAN^{-1}(x)$, tanh(x). **#1)**
EXPT	Exponentiation of a variable with another variable. a^b = EXPT(a,b) = 2^5 = 32

All the above functions are normally built-in in a PLC; i.e. functions which can be used without adding an extra programming library (add-ons) or program modules. Small variations in the functions can occur in relation to every PLC-type. It is recommendable always to look through the manual from each individual PLC-type in order to gain an overview and see the possibilities for math functions and routines.

Remember to check the variable data types for each individual math function, in order to use the right one.

#1) To calculate between radians (RAD) and degrees (GRAD) see page 49.

7.4   Logical Operators (AND, OR, XOR, NOT)

Logic operators are used to compare two different BOOL variables or values. The result of the comparison is a value, which always has the data type Boolean (BOOL) and can thus only be TRUE or FALSE.

The possible operators including examples are as follows:

Operator	Description	Example S1:= TRUE, S2:= FALSE S3:= TRUE	Result
&	Same as AND, Only TRUE if both values are TRUE	K1:= S1 & S2 K2:= S1 & S3	K1 = FALSE K2 = TRUE
AND	AND, Result is TRUE if both values are TRUE	K1:= S1 AND S3 K2:= S1 AND S2	K1 = TRUE K2 = FALSE
OR	OR. TRUE if one value is TRUE	K1:= S1 OR S2 K2:= S1 OR S3	K1 = TRUE K2 = TRUE
XOR	The result is TRUE is values not are equal.	K1:= S1 XOR S2 K2:= S1 XOR S3	K1 = TRUE K2 = FALSE
NOT	not, TRUE result in FALSE FALSE result in TRUE	K1:= S1 AND NOT S2 K2:= NOT S1 K3:= NOT S2	K1 = TRUE K2 = FALSE K3 = TRUE

Logic is mostly used together with IF-statements, as described on page 53.

AND can be used in serial-connected components (sensors/contactors/switches), where all components must give ON signal to make the total expression be TRUE. OR can be used in parallel-connected components, where just one component must give ON signal, to make the total expression be TRUE.

The logical operators can also be used directly on e.g. binary values as shown:

```
Var1 := 2#10010011 AND 2#10001010;   // Var1 = 2#10000010
```

```
Var2 := Var1 OR 2#10001010;   // Var2 = 2#10001010, DEC138
```

7.5 Logic, math formulas and parentheses ()

It is important to be aware of how math formulas are calculated in a PLC Controller. If you are in doubt of how values are calculated – is the addition the first action to perform or is it the multiplication the first action in a formula – it all depends on where to place the parentheses.

> The mathematical rules tell that *multiplying* is carried out before *plus*, but experience shows you cannot be 100 % sure that the rules of math are respected correct in a PLC or that the formula is written correctly in the PLC Code. Then use parentheses

If the math formula contains Boolean expressions like **AND** or **OR** as shown below:

```
X:= B1 OR B2 AND B3;
```

Then **AND** is read as 'multiply' and is calculated first. **OR** is read as 'plus'.

It means: if the value of **B2** is **FALSE**, then the expression **B2 AND B3** is **FALSE**.

If you are in doubt of the result, then use parentheses as shown below:

```
X:= B1 OR (B2 AND B3);
```

The next example is this formula:

$$V1 = \frac{V2}{V3} + \sqrt{(V4 + V5)}$$

The formula can be written in the PLC code having extra parentheses as follows:

```
V1:= (V2/V3) + (SQRT(V4 + V5));
```

SQRT is the mathematical function in a PLC calculating a square root. The function has only one input parameter. **V4** and **V5** are added, before calling the function.

8 Working on variable assignment

Variables is a central part of programming. In this chapter, some of the basics will be shown and what should be taken into consideration when working with variables. Variables are in some PLC types called **tags** or **PLC tags**.

> **Definition:** A variable points at a box in the memory containing a place, in which a numerical value can be written. The size of the box depends on the data type which is very important to remember.

Below it is shown that a variable with the name of **VarA** get a copy of the number which exists in the variable **VarB**.
Notice the use of the signs := and ; (Colon, equals and semi-colon)

> **VarA:= VarB;**

Subsequently **VarB** can be given a value of 17.6 as follows:

> **VarB:= 17.6;**

Dot (.) is always used in a PLC when decimal number are applied. Both the variable **VarB** and **VarA** has the data type **REAL** (**REAL** is used for decimal digits).

If the data type for **VarB** is an **INT** (integer) it is normal that the compiler (the program in which the PLC code is written) comes up with a warning that data will be lost, as the number, which is attempted to be written in **VarB** is a decimal number (17.6). This is due to the fact that the variable **VarB** only can contain a integer, if it is created with the data type **INT** (integer)

When working on variables, the calculations are simple to work with in ST-programming. This is a calculation, written directly in a PLC code:

> **VarB:= 17.6 * 8 + VarA;**

If the contents of the variable **VarA** is 23, then the contents of **VarB** is 163.8

Below the variable **Count** is shown, which at each program-scan is increased with 1 (1 is added to the previous value). The program execution has an internal variable for calculations (called Stack/Accumulator) and it makes a copy of the variable **Count**, adds 1 to it and returns the new value to **Count** again:

```
Count:= Count + 1;
```

If **Count** is of the data type INT, you must be aware that when **Count** reaches the value 32767, then it will change to – 32768, the next time (next program scan). It is the programmer's task to make sure that no overrun happens on a variable. There are two methods to do this: Either a larger variable is used for **Count** e.g. DINT Or the counter is reduced with a condition (IF-statement, se page 53), setting the value to 0, when the number is large.
The last method is the best one, as no overrun happens to the variable:

```
Count:= Count + 1;

IF Count > 99 THEN //Check value. Must be below 99
   Count:= 0; //Reset counter
END_IF;
```

As shown above, **Count** adds 1 for each program-scan. If the PLC scan time is set to 1 [ms] it will take 100 [ms] before **Count** is reset to 0.

TIP: The above counter can easily be used as a program *Heartbeat*, thus it is possible to see an activity on the current PLC program.

The following built-in counting functions exist: CTU, CTD and CTUD. See page 88.

8.1   MATH calculations challenge

Math and calculations, where formulas are involved, are easily to work with in ST-programming. It is one of the biggest advantages compared to the other PLC pro-gramming languages. However, there are many things which should be taken into consideration when working with math functions and formulas. These are:

- Division by 0 (chapter 8.2, page 46)
- Calculating with INT and REAL (chapter 8.3, page 47)
- Decimal errors with REAL (chapter 8.4, page 48)

8.2   Division by zero

A PLC reads data from different sensors and they can be zero without being thought of. A temperature meter e.g. in an office showing 20 degrees Centigrade, but if the temperature drops to zero when outdoor, it goes wrong. This can be shown in the below calculation, where **VarC** is equal to **VarA** divided by **Temperature**:

```
VarC:= VarA / Temperature;
```

If **Temperature** becomes zero, the PLC will receive a run time error and/or become unstable as it is an invalid mathematical operation in a PLC.

To ensure that the PLC does not receive a run time error at any time and minimize the risk of errors occurring later, the above PLC code can be changed as follows:

```
IF Temperature <> 0 THEN
   VarC:= VarA / Temperature;
END_IF;
```

The calculation is thus only carried out if **Temperature** is not zero (The operator sign **<>** means different from / not equal. See chapter 7.2, page 39

Another possibility to ensure that the calculation is not carried out when the **Temperature** is zero, is the following solution:

```
//Insure temperature is not zero when calculating
IF Temperature = 0 THEN
   Temperature:= 0.0001;
END_IF;

VarC := VarA / Temperature;
```

NOTICE: These math functions cannot tolerate that **x** is zero: LN (x) and LOG (x).

8.3 Calculating with REAL and INT

Calculations can be made with both integers (INT) and decimal values (REAL). If the average of two integers is to be found it must be considered which data types the variable are using and how the calculation is carried out. In the below example all variables are of the type INT (the brown box shows the value of the variable):

```
24    varA   10   :=10;
25    varB   15   :=15;
26    VarC    0   :=VarA   10  /VarB   15  ;
```

The result is that **VarC** will be zero, as **VarC** is an integer. The calculation shown is a division between two integers, resulting in a decimal value (10 divided by 15 does not result in an integer, but 0.67). In order to make the calculation succeed, the calculation have to takes place in a REAL. The calculation in a PLC is made by using the data type for the calculation which the first variable in the formula provides, namely the data type for **VarA**. When calculating, it is not important that **VarC** is a REAL. However, **VarC** must be a REAL, because otherwise the result cannot be saved (It is not possible to save a REAL in an INT).

The solution is that **VarA** is copied to **VarC**, which is of the REAL type. The calculation takes, therefore, place internally in a REAL variable. The code must be as shown:

```
    varA   10   :=10;
    varB   15   :=15;
    VarC   0.667  ▶  :=VarA   10  ;
    VarC   0.667  ▶  :=VarC   0.667  ▶ /VarB   15  ;
```

A tip is therefore to check whether the calculation shows the expected result, as calculations in a PLC can be incorrect in relation to what is known from our calculators. A calculator or a math program must be used to make control calculations.

In some PLC types, a calculation is made in an accumulator (ACC), where values must be copied to and from. However, the same rules apply.

8.4   Decimal errors by REAL

When calculations are made with REAL it can be experienced that a value is not a round number. It might be expected that the value for a variable is a round nice number as e.g. 11 but the number is 10.999999. This is caused by the fact that a computer can only work with integers and a REAL value, is an adjusted value. This can create a problem when comparing numbers. This is shown in the below, where a variable **Lamp1** must be set to 1, when the variable **Sensor1** becomes 11:

```
IF Sensor1 = 11 THEN
   Lamp1:= 1;
END_IF;
```

There is no guarantee that the above PLC code will work efficiently – that is that **Lamp1** is set to 1, as a decimal error may occur which has the consequence of **Sensor1** never become exactly 11. The above example must thus be changed to the below PLC code, where **Sensor1** must be placed in a certain range rather than at single specific value. The range could be between 10.99 and 11.01:

```
IF (Sensor1 > 10.99) AND (Sensor1 < 11.01) THEN
   Lamp1:= 1;
END_IF;
```

As an alternative, use the rounding off functions **FLOOR**() or **TRUNC**(). See page 40.

It is possible to implement a rounding off function (here rounding off is to 1 decimal):

1)   Multiply the sensor value by 10
2)   Convert value to an INT variable with the function: REAL_TO_INT();
3)   Convert back to a REAL variable with the function INT_TO_REAL();
4)   Divide value by 10

Problems when rounding can e.g. be experienced when a motor is not running totally idle as the speed is not exactly 0 (zero) or a tank is physically empty, but the level sensor shows a small value close to zero. A flow meter can also show a small value even if the facility is not in operation. This can, however, be caused by lacking calibration (zero position) of the instrument.

8.5   Data communication variables

There is often a need to transfer variables to other computers, which are parts of the total automation solution. This chapter describes problems which needs attention in connection with data communication.

Problems can occur when transferring REAL variables to other PLCs, PCs, electrical apparatuses or automation instruments. This happens due to many different interpretations of how a REAL or FLOAT value is defined in different computers. (A computer only works in integers). This can also happen due to different programming versions or that 16, 32, 64, 128-bit systems understand REAL and FLOAT differently. This problem is solved by always transferring values in data communication as integers. Values can then be multiplied by 100 to obtain values with two decimals and the receiver must then divide by 100 to obtain the right decimal values with two decimals.

It can be a challenge to transfer STRINGS between computers. This can happen due to different ways of handling and interpretations of STRINGS. They could be different bit sizes, Unicode, the choice of ASCIIs which create the challenges. Furthermore, the length of a STRING is indicated at position zero in some programming languages. If STRING is converted to BYTE, then it is 'simple' to transfer data.

Always start by obtaining a data communication by reading WORD. Remember that some computers have swapped WORD values (the lowest 8 bits are exchanged with the 8 highest bits) and remember that if value begins with **0X**, it is a HEX value.

In some PLCs, a BOOL fills up 16 bits and can consequently also be an INT.

8.6   Data type conversations functions

If the content of a variable with only one data type is needed to be transferred to a variable with another data type, a lot of built-in functions can be used. Some PLC types have more than 100 different converting functions for the different data types. Naming and size of the functions are as follows:

Type1_TO_Type2 (**ConvertFrom**);

where

Type1 is the copy from data type (Data type of **ConvertFrom**).
Type2 is the converted data type

Table with chosen datatype converting functions:

Function	From	to	Example	Comments
REAL_TO_INT	REAL	INT	Val:= REAL_TO_INT(1.6); \\Val = 2 Val:= REAL_TO_INT(1.3); \\Val = 1	Rounding to nearest integer (IEC60559) **Val** is an **INT**
INT_TO_REAL	INT	REAL	Val1:= INT_TO_REAL(4); \\Val1 = 4.0	Val1 is a **REAL**
INT_TO_BOOL	INT	BOOL	Val2:=INT_TO_BOOL (1); \\Val2 = TRUE Val2:=INT_TO_BOOL (0); \\Val2 = FALSE	1 is converted to TRUE. 0 is converted to FALSE.
INT_TO_TIME	INT	TIME	Val3:=INT_TO_TIME (5); \\Val3= T#5ms Val3:=INT_TO_TIME (60); \\Val3= T#60ms	Converts a integer value to a variable with the TIME data type with the resolution in [ms] TIME can only be converted to a integer value because TIME is a counter value which counts from 00:00:00 UTC #1)
RAD_TO_DEG DEG_TO_RAD	LREAL	LREAL		Converts between radians (RAD) and degrees (GRAD). Used together with the SIN and COS functions

The function **REAL_TO_INT** must be used, if a conversion from a **REAL** variable (decimal value) to an **INT** variable is to be made. Se first row in the above table.

It is important to secure that the value *can* be converted, as an error can occur, causing the PLC to stop the program execution or making the program unstable.

#1) DATE is converted from an internal electronic circuit, which is a part of the hardware in a PLC. This circuit is counting time in seconds from 00:00:00 UTC 1.1. 1970 (Coordinated universal time, atomic clock). Notice that the next Y2K appears in year 2038.

8.7   Find binary values of an integer (Masking bit)

In some cases there is a need to convert an integer value into a binary value, and thus control whether a specific bit in a variable is set **TRUE**. Typically used when different digitals output (e.g. lamps) is set from an integer value.
This feature is also called: Mask out the binary digit from an integer.

This can be carried out in a simple way: Use dot and a digit (bit position no. 0) after the variable as shown below:

```
MyUINT:= 3;     //Unsigned INT datatype. The BIN value is 2#0011
MyBOOL2 := myUNIT.0; //Get bit 0 from MyUNIT value
```

MyBOOL2 (BOOL data type) is **TRUE**, because position 0 in **MyUINT** is the first bit, which is 1 in a value that is 3.

The above can also be written as follows:

```
MyUINT4:= MyUINT AND 2#001;  //Where MyUNIT = 3 = 2#0011
MyBOOL:= UINT_TO_BOOL (MyUINT4); // Convert to a BOOL
//Result is that MyBOOL is TRUE
```

Where **AND** can be used to mask out a bit at position no. 0. Each bit in the two values **MyUINT** and **2#001** are 'multiplied binary', and if the result is 1, the final result will be **TRUE**. When '2#' is placed before a value, it means that the value must be interpreted as a binary digit. See chapter 4.1, page 14.
If there is a need for a result that must be a variable with a BOOL data type, the converting function UINT_TO_BOOL must be used.

Below is an example where a variable named **Var1** (UINT) is used to set different outputs bits. The variable **OutPutBit** will set to **TRUE**, if a binary value is 1 in **Var1**:

```
OutPutBit1:= UINT_TO_BOOL(Var1 AND 2#00001); //Set bit if bit 0 is TRUE
OutPutBit2:= UINT_TO_BOOL(Var1 AND 2#00010); //Set bit if bit 1 is TRUE
OutPutBit3:= UINT_TO_BOOL(Var1 AND 2#00100); //Set bit if bit 2 is TRUE
```

8.8   Convert REAL into 2 decimals (2 digit REAL)

If a REAL value is converted into a STRING and read out on HMI (user Interface) or written to an ACSII file, the value will often include 7 to 9 digits. That many digits are not very readable and user-friendly. It is, however, a way for a computer to handle a decimal digit. A LREAL data type has 15 digits.

The method below converts the value in **RealNumber** to a digit with 2 decimals. If 3 decimals are required the constant **DecimalFactor** must be 1000:

```
VAR CONSTANT
    DecimalFactor : REAL := 100; //10 for 1 digits, 100 for 2 digits, 1000 for 3 digits
    RealNumberBegin : REAL := 50.7172;
END_VAR
VAR
    INTNumber: INT;   RealNumber: REAL ;
END_VAR

RealNumber:= RealNumberBegin;
IF DecimalFactor > 0 THEN //Avoid division by zero (0)
    RealNumber:= (RealNumber * DecimalFactor) + 0.5; //+ 0.5 to round up   #1)
    INTNumber:= REAL_TO_INT(RealNumber);      //  Convert to integer  #2)
    RealNumber:= INT_TO_REAL(INTNumber);      //  Convert to decimal #3)
    RealNumber:= RealNumber/DecimalFactor;    //  Add decimal        #4)
END_IF;
```

DecimalFactor is a CONSTANT, because is used more than once in the PLC code.

A calculation example, where 50.7175 is to be converted to 50.72:

#1) (50.7175 * 100) + 0.5 = 5072.25
#2) REAL_TO_INT (5072.25) = 5072 (Integer value)
#3) INT_TO__REAL (5072) = 5072 (decimal value)
#4) 5072/100 = **50.72**

IMPORTANT
Rounding off must not be carried out before other calculations, as it deletes information. Rounding off must only be carried out, if value is to be shown to the user:

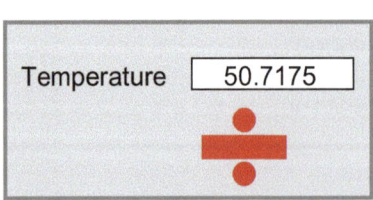

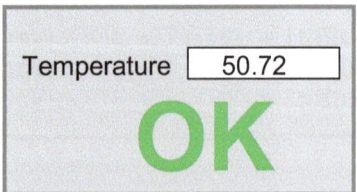

9   Conditional Statement

The following chapter consists of the central statement concepts in ST.

In the general format descriptions, **<Condition>** and **<Statement>** must be replaced by variables, expressions and PLC code.

9.1   IF-THEN-ELSE

An IF-THEN-ELSE statement – or a sentence – is the most used expression in ST programming.

An IF statement can e.g. be used to control whether a digital sensor shows a signal (e.g. an electrical start contact, an ON/OFF switch or a level contact in a pump well). If the digital sensor shows a signal, an action is to be taken, being e.g. starting a pump or switch on a lamp. An IF statement can be used both when using analogue and digital input-signals. Further, the IF statement can be used for internal variables.

The general format of the IF statement is as follows:

```
IF <Condition> THEN
    <Statement>
END_IF;
```

Where:

<Statement> =	Can contain one or more lines of PLC code, always ended by END_IF and semicolon.
<Condition> =	An expression, always being TRUE or FALSE. If the expression is true, the PLC code in <Statement> is carried out.

The <Condition> line can e.g. be an input signal from an electrical contact or sensor and the <Statement> line can be an output signal to switching on or off a lamp.

The statement ELSE can be added to the expression:

```
IF <Condition> THEN
   <Statement>
ELSE
   <Statement1>
END_IF;
```

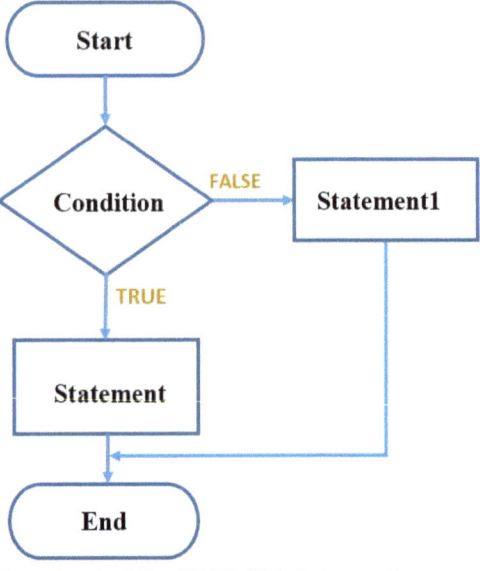

As can be seen, the ELSE part is optional and notice that the lines including **<Statement>** is tabulated (2 x blanks) to make the whole expression more readable.

The mode of operation is as follows:

If **<condition>** is fulfilled (**TRUE**), the PLC code in **<Statement>** will be carried out.

If **<condition>** is *not* fulfilled (**FALSE**), the PLC code in **<Statement1>** is carried out.

Flowchart of the IF-ELSE statement

NOTICE: If **<Condition>** section contains colon ":", it indicates that it is controlled if the variable assignment is carried out well, which is not normally the intention! Therefore, remember that the sign "=" must stand alone as shown below:

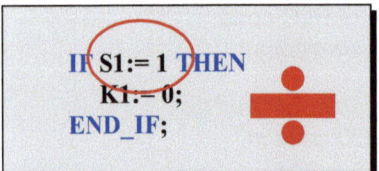

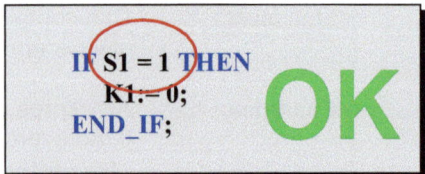

It is possible to create the statement more complex, as shown below:

```
IF <Condition1> THEN
   <Statement1>
  IF <Condition2> THEN
     <Statement2>
  ELSE //<Condition2>
     <Statement3>
  END_IF; //End the <Condition2>
ELSE
   <Statement4>
END_IF; // End the <Condition1>
```

The mode of operation is as follows:

The PLC code in **<Statement1>** is executed, if **<Condition1>** is TRUE. After **<Statement1>** has been carried out, **<Condition2>** is controlled and if it is TRUE, **<Statement2>** is carried out; otherwise **<Statement3>** is carried out. If **<Condition1>** is FALSE, then **<Statement4>** is carried out instead.

Take care, because it will soon be to complex with many ELSE!

IMPORTANT
If there are many (more than 3) IF-THEN-ELSE statements, the PLC code can be difficult to read. A CASE statement (chapter 9.2, see page 58) can easy replace complex IF statements in order to increase the readability of the code.

It also minimizes the possibility of making errors in complex IF-THEN-ELSE statements, when other people are to correct or add something in the PLC code.

Furthermore, the amount of lines in the PLC code is reduced, when many identical IF statements are replaced with a CASE. A reduction of more than 50 % in the amount of PLC code lines is not unusual, when CASE is used. See page 61.

It is possible to add an ELSIF statement to control more conditions:

```
IF <Condition1> THEN
   <Statement1>
   ELSIF <Condition2> THEN
      <Statement2>
      ELSE
      <Statement3>
END_IF;
```

The mode of operation is as follows:

The PLC code in **<Statement1>** will be carried out, if **<Condition1>** is TRUE. If **<Condition1>** is FALSE the **<Condition2>** is controlled, and if it is TRUE the PLC code in **<Statement2>** will be carried out. If not **<Condition1>** or **< Condition2>** are fulfilled, the PLC code in **<Statement3>** will be carried out.

It is recommended to use CASE statements instead of ELSIF.

9.1.1   EXAMPLE: IF-THEN-ELSE as latching relay

This example shows a motor controlled by a latching relay. A latching relay maintains its state after being activated. Also called *keep*, or *stay* relay.
There is a switch-on contact with a variable named **B1Start**, which has the data type BOOL and it is a **N**ormally **O**pen (NO) contact. Furthermore, there is a stop contact named **B1Stop** with the data type BOOL and it is a **N**ormally **C**lose (NC):

```
IF  B1Start THEN        //Normally  Open (NO) contact
   K1Motor:= TRUE;      //run motor
END_IF;

IF  NOT B1Stop THEN  //Normally Close (NC) contact
   K1Motor:= FALSE;  //Stop Motor
END_IF;
```

When the **B1Start** contact is activated, the **K1Motor** is set to TRUE and the motor starts. The **K1Motor** is fixed as TRUE even if **B1Start** is not activated subsequently. The data type for the **K1Motor** variable is BOOL. If the **B1Stop** contact is activated, the **K1Motor** will be set to FALSE and the motor stops. NOT is written before **B1Stop** as the **B1Stop** signal is physically short-circuited in the electrical contact and is therefore normally TRUE (same as positive signal) on the digital input module.

9.1.2 EXAMPLE: IF-THEN-ELSE open and close valve

The following PLC code example checks the alarm from a pump and the pressure in relation to a set-point:

```
IF ((PumpAlarm = TRUE) AND (PumpPressure > PumpSetPoint)) THEN
    PumpValveOpen := TRUE;   //Open value
ELSE
    PumpValveOpen := FALSE; //Close valve
END_IF;
```

If the whole condition, - marked by the outer parentheses in the IF statement – is **TRUE**, the value **PumpValveOpen** is opened, otherwise the valve is closed.
PumpAlarm can be a digital input with the data type BOOL. **PumpPressure** is a variable with the data type REAL and can be of a value which the user can adjust, e.g. via a user control panel (HMI).

The above PLC code example can be rewritten to:

```
PumpValveOpen := FALSE;   //#1 Note

IF (PumpAlarm = TRUE) AND (PumpPressure > PumpSetPoint) THEN
    PumpValveOpen := TRUE;   //#2 Note
END_IF;
```

This means a line less and is simpler to read for some programmers.

NOTICE: Values are only moved to the output modules, when *all* PLC codes are executed (a program scan), therefore the connected valve will not close (see **#1**) and open **#2**) instantly again.

In order to make the PLC code simpler, it can be rewritten as follows:

```
PumpValveOpen :=  PumpAlarm AND (PumpPressure > PumpSetPoint);
```

The variable **ValveOpen** is set to **TRUE** or **FALSE**, without using an IF statement!

9.2   CASE

CASE is a statement used when different events are to be carried out based on only one variable. CASE is to be used, when IF statements become too complex. CASE is a really good for sequence control (state machine) and is often applied, when e.g. a machine is positioned in different operational modes (e.g. STOP, START-ING, RUN, STOPPING) or applied in a process in a dairy (e.g. NONE, CREAM, SKIM_MILK, WHOLE_MILK, WATER_FLUSH).

A CASE statement has the following format:

```
CASE <Condition> OF
  <SelectorValue1> : <Statement1>;
  <SelectorValue2> : <Statement2>;
  <SelectorValue3> : <Statement3>;
  ...
ELSE
  <SelectorValueELSE>
END_CASE;
```

Mode of operation:

The variable determining the event is **<Condition>** and this must be an integer type.

The different values, **<Condition>** can take on, are written in the sections **<Selector-Value1>**, **<SelectorValue2>** and **<SelectorValue3>** followed by colon ":"
This to be carried out is written in **<Statement X>**, here **X** is 1, 2 or 3 and it can be PLC code. If this PLC code is longer than 4 to 6 lines, a function or a program module should to be created, in order to have a readable code.

If **<Condition> = <SelectorValue2>** the code in **< Statement2>** will be executed.

It is not required to have PLC code in **<Statement>**. The sections can be empty.

The tree dots/points (...) indicate that the amount of PLC code lines can be free of choice; however, at least one line.

The ELSE section is free of choice. It is, however, recommendable that some PLC code are written in this section, as e.g. an alarm message or a note of errors, so that the programmer is aware that a program call is performed in the ELSE section.

9.2.1   EXAMPLE: CASE – Setting the motor speed

Here is an example of how to use a CASE, where the speed of a motor is adjusted on an electrical switch named **MotorSwitch**. The switch can be turned in steps from 1 to 6 which can be 6 voltage levels. **MotorSwitch** is an INT data type.

```
MotorFan:= 0;  //Turn off the motor cooling

CASE MotorSwitch OF
  1, 2 : MotorSpeed := 25;   //Two values in CASE, separated by comma
  3    : MotorSpeed := 35;   //One value in CASE
  4..6 : MotorSpeed := 50;   //Interval CASE: start no. .. end no
         MotorFan:= 1;       //Turn on the motor cooling
ELSE
  MotorSpeed := 0;  //Use as default
END_CASE;
```

Explanation of this example:

If **MotorSwitch** is 1 or 2, the **MotorSpeed** will be 25. If **MotorSwitch** is 3, the **Motor Speed** will be 35. If **MotorSwitch** is 4, 5 and 6, the **MotorSpeed** will be 50.
If no CASE is fulfilled, i.e. **MotorSwitch** is not 1 to 6, the **MotorSpeed** will be zero.

When the **MotorFan** variable is always set to 0 (cooling off) before the CASE code begins, it is easy to set the **MotorFan** to 1 in the CASE where the cooling have to run. This avoids having to put the **MotorFan**:= 0 lines in all other CASE sections.

As can be seen in the example, the ELSE secures that **MotorSpeed** will be zero (sets the motor speed to zero), if **MotorSwitch** has a value which the CASE does not know, - it secures that the PLC code has a better quality and a decision is taken, on what is to happen when an unknown value of **MotorSwitch** occurs.

It is recommended to replace the values 1, 2, 3, 4 and 6 with variables, created as CONSTANT, as the values occur in many positions in the same PLC code, because there is a risk that the PLC programmer 'forgets' to change all the positions in the PLC code, if it is necessary to change the values. Read more about CONSTANT in chapter 6.2, page 36 and **ENUM** in chapter 4.4, page 21.

9.2.2   EXAMPLE: CASE – For executing programs

This chapter describes an example, where CASE is used to execute different program modules. See more about division/split up in program modules in chapter 10, page 68. The example is shown without and with using CONSTANT:

The value of **ProgramSelect** determines which program module has to be executed:

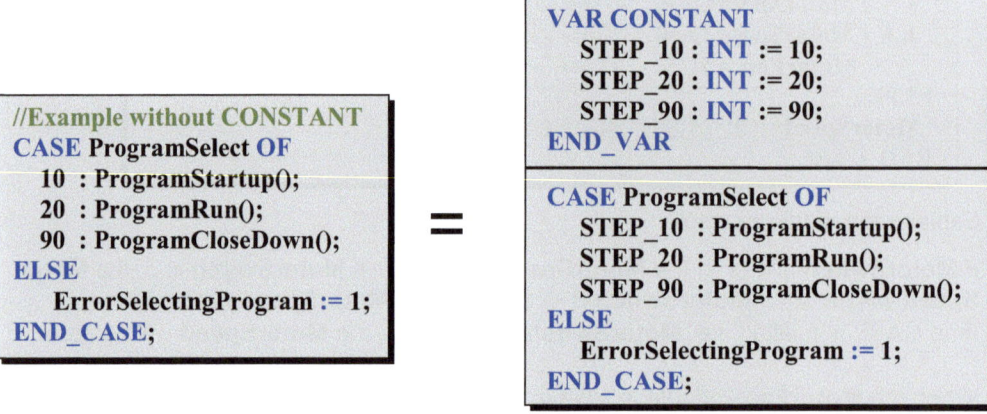

Mode of operation:

If **ProgramSelect** is 10, then the program module **ProgramStartUp** is executed. Inside the **ProgramStartUp**, the variable **ProgramSelect** is changed to be 20, so that **ProgramRun** is executed in the next program-scan instead of **ProgramStartUp**.

If the condition **ProgramSelect** is set to a value which is not implemented in CASE the variable **ErrorSelectingProgram** is set to 1, so that the programmer can be informed that a program module is not chosen.

The fixed values, which **ProgramSelect** can be, are: 10, 20 or 90. A jump between the values (11, 12, 13 .. to 19 are free) is made on purpose in order to give room for future extensions. CONSTANT can advantageously be used instead of fixed values. Read more about this in chapter 6.2, page 36

IMPORTANT   A solid software structure is created by using CASE to execute different program modules. CASE gives a better overview than many IF-statements and especially ELSE-IF statements.

9.2.3   EXAMPLE: CASE – Recognizing numbers

In this chapter, an example is showing how a CASE statement is used to recognize numbers. The numbers could be a range of certain passwords, which are to be recognizable so that a user can gain access to the control panel (HMI). There are often various levels for gaining access as e.g.:

>Operator Password
>Administrator Password
>SuperUser Password

In the following example a variable **PassOK** is used for determining whether **PassSelect** contains a valid password. Firstly, the variable **PassOk** is a BOOL and is set to FALSE.
If the variable **PassSelect** contain one of the values 1747, 3309, 5607, 1234 or 1027 the variable **PassOK** is set to TRUE as shown:

```
PassOk:= FALSE; //No valid password number

//Check password
CASE PassSelect OF
   1747, 3309, 5607, 1234, 1027: PassOk:= TRUE; //Valid pass number
END_CASE;
```

The example shows that CASE is a simpler solution than applying many IF statements, because the above solution will require five IF statements (15 lines of PLC codes) or a very long IF statement, as shown in the example below:

```
PassOK:= FALSE;

//Check password
IF PassSelect = 1747 OR PassSelect = 3309 OR PassSelect = 5607 OR
      PassSelect = 1234 OR PassSelect = 1027 THEN
   PassOK:= TRUE; // Valid password number
END_IF;
```

As can be seen above, long lines can be written, but the PLC code will consequently be difficult to read. Further, It is recommended that the PLC code lines not are longer than the screen width of the PLC compiler programming tool.

9.3   Iteration statement, LOOPS

Loops are used when a PLC code is to be repeated a number of times. Loops are often used when all values in an ARRAY must have a certain value or a maximum or minimum value must be found in an ARRAY.

In chapter 9.4.3 found on page 66, an example is showing how to find an average value in an ARRAY.

It is important to secure that no DEAD LOCK occurs in the PLC, which is a situation where the CPU uses all its power to work on the loop and this is a typical error. In order to secure that the loop is always ends, it must end after a maximum time or a maximum amounts of executions.

The following chapter shows different methods of implementing loops.

9.4   FOR-DO Statement

This type of loop is the most frequently used type. A FOR-DO is always executed a certain amount of times. It is determined by a start value and an end value.

Format is as shown below:

```
FOR <StartValue> TO <EndValue> DO
  <Statement>
END_FOR;
```

Where:

<StartValue> =	A countable variable (INT or DINT) having the start value.
<EndValue> =	Execute the **<Statement>** until the countable variable reach this value
<Statement> =	Containing the PLC code to be executed at each execution. It can be one or more PLC code lines.
	It is recommended that lines between FOR and END_FOR starts with 2 X SPACE, because it is more readable.

NOTICE   It is not allowed to change the counting variable in the **<Statement>** section – it interrupts the program execution!

A loop always counts 1 ahead each time. If there is a need to count more than 1 ahead at the time, BY is added. However, not used very often.
If the loop needs to step backwards (**StartValue** > **EndValue**) use BY -1.

Format using **BY**:

```
FOR <StartValue> TO <EndValue> BY <StepValue> DO
   <Statement>
END_FOR;
```

NOTICE:
The smaller PLC types, not having that much calculation capacity, cannot handle large FOR-DO statements that has too large scanning times. In this case, it is recommended to reduce the FOR-DO statements or to split up the FOR-DO statements in minor FOR-DO statements and place these in different program modules and execute them with different scan-times.

IMPORTANT
A typically error when working with FOR-DO statements and ARRAYs is that the first or last position in the ARRAY are not handled. Further, the loop is running longer than the size of the ARRAY which perhaps results in unstable PLC code.

Variable names such as i, j, n or m are often used as countable variables.

If there is a need of exiting from the FOR-DO loop before all executions are carried out, it is possible by adding the EXIT command. A certain value might be looked for in an ARRAY and when this value is found, it is not logical to continue the loop.

Format with EXIT:

```
FOR <StartValue> TO <EndValue> DO
   <Statement>
   IF <Condition> THEN //#1, Exit now?
      EXIT;          //Exit the loop
   END_IF;
END_FOR;
```

As shown above, an IF must be added (#1) within the FOR-DO statement and if the **<Condition>** is TRUE, the EXIT will be carried out and the loop immediately ended.

9.4.1   EXAMPLE: FOR – LOOPS, 4 times loop

In this chapter an example shows where an ARRAY is created with 4 elements, all having the data type INT, and each of the elements is set to 7 by using a FOR-loop:

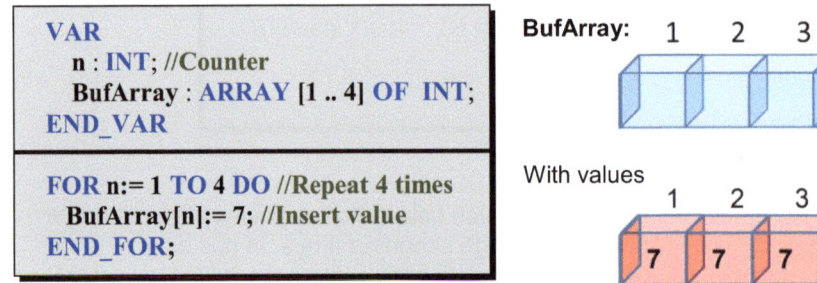

As can be seen, the number 1 and 4 occur twice in the example, when ARRAY is created and in the FOR-loop. These numbers must, therefore, be created as a CON-STANT, because a typical error is the fact that the programmer forgets to correct in both positions in the PLC code.

The FOR-loop in the example above can be rewritten to the following four lines:

```
BufArray[1] := 7;
BufArray[2] := 7;
BufArray[3] := 7;
BufArray[4] := 7;
```

Here it can be seen that the counter variable **n**, used by the FOR-loop, counts 1 ahead each time the loop makes one execution. In each execution, the value 7 is inserted on the position in **BufArray**, of the relevant variable n.

As seen, the FOR-loop replaces four lines of the PLC code!

Single values can be inserted directly in **BufArray** as follows:

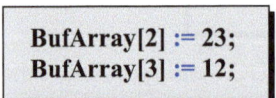

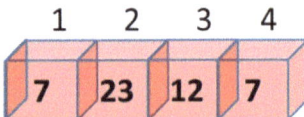

9.4.2   EXAMPLE: FOR – LOOP and 3D ARRAY

The example in this chapter shows how all elements in a 3-dimensional ARRAY named **Array3D** are set to 1. This method can be used right after the program starts, or if all positions in an ARRAY must have a certain value.

In the variable section three supporting variables are created: x, y and z, which are used as indexing in the ARRAY.

In order to define the size of the **Array3D**, three CONSTANT variables are created: **X_MAX**, **Y_MAX** and **Z_MAX**, so it is simple to change the size of the ARRAY later. ARRAY might have another size during test and the structure process of the PLC code, and by using CONSTANT all positions are changed.

```
PROGRAM MAIN
VAR CONSTANT
    X_MAX : INT := 10;
    Y_MAX : INT := 20;
    Z_MAX : INT := 30;
END_VAR
VAR
    x, y, z : INT; //Index to the 3D Array
    Array3D : ARRAY [1 .. X_MAX, 1 .. Y_MAX, 1 .. Z_MAX] OF INT ;
END_VAR

FOR x:= 1 TO X_MAX DO
   FOR y:= 1 TO Y_MAX DO
      FOR z:= 1 TO Z_MAX DO
          Array3D[x, y, z] := 1;      //Set current position to 1
      END_FOR;
    END_FOR;
 END_FOR;
```

A 3D ARRAY can e.g. be used as follows: Placing packages on a pallet in a production line, a large warehouse logistics terminal or a big parking garage.

In the above example, 10 x 20 x 30 = 6000 elements with INT variables are created. It can result in execution problems on smaller PLC types, when a loop with 6000 elements are executed. If this is the fact, a number of 2D ARRAY can be created instead, as a 3D ARRAY can always be rewritten to an amount of 2D ARRAY.

9.4.3 EXAMPLE: Calculation of the average value

The following example shows how a FOR-DO loop can be used to calculate an average value of a range of values saved in an ARRAY. It is assumed that the values, which need an average calculation, are already saved in **BufArray**:

```
PROGRAM Average
VAR CONSTANT
    BufArrayMin : INT := 0;
    BufArrayMax :INT := 9; //Must be higher than BufArrayMin
END_VAR
VAR
    i                    : INT;    //Counter variable in FOR-DO statement
    BufArray             : ARRAY [BufArrayMin .. BufArrayMax] OF REAL;
    BufArrayTotalSum : REAL;   //Calculator for the total value sum
    BufArrayAverage   : REAL;  //Average value of the BufArray
END_VAR

BufArrayTotalSum := 0; //Reset calculator #1)

//Sum all values from the buffer into BufTempVar #2)
FOR i := BufArrayMin TO BufArrayMax DO
   BufArrayTotalSum := BufArrayTotalSum + BufArray[i];
END_FOR;

//Calculate average
BufArrayAverage := BufArrayTotalSum /( BufArrayMax – BufArrayMin + 1 );
```

Overview of ARRAY having 10 positions (elements):

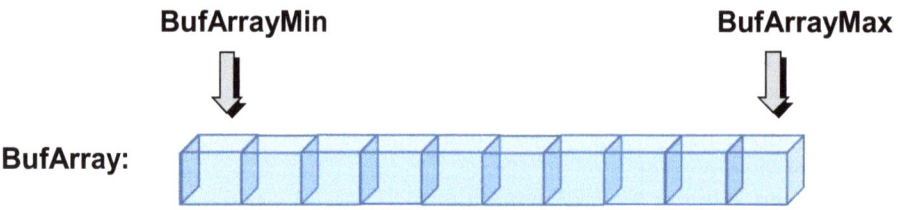

BufArrayMin

BufArrayMax

BufArray:

Explanation to the **Average** example program:

Constants

Two constants, **BufArrayMin** and **BufArrayMax**, are created, because as the constants are used three times in the PLC code and when changing the length of the **BufArray** it is certain that all constants are changed.

Naming

Constants and ARRAY have the same first name, **BufArray**, indicating that they belong to each other.

Mode of operation

BufArrayTotalSum is a variable containing the total sum and is created with the data type REAL. Firstly, the variable **BufArrayTotalSum** is initialized to the value zero (0) to secure that the content is zero first time running. #1)

The next step is that all values in the **BufArray** are added together by using a FOR-DO loop. The **BufArray**, which **BufArrayMin** is pointing at, is at the first position, and the ending **BufArray** is the value which **BufArryMax** is pointing at. Notice that the times the FOR-DO loop executes is **BufArrayMax – BufArrayMin + 1** times as the first and the last execution are both included. #2)

When the loop is finish the variable **BufArrayTotalSum** now contains all values, added together. In order to find the average value, the total amount is divided by the number of the times the FOR-DO loop has executed. The result is consequently placed in **BufArrayAverage**.

It is important to make sure that the result of the calculation **BufArrayMax – BufArrayMin + 1** time is not zero, because a PLC cannot handle a division by zero.

The calculation of the average value in a PLC is often used to filter input signals from analogue sensors. When filter the signals, 'noise' might be removed from the measuring. The disadvantage of a FOR-DO loop for this purpose is that ARRAY needs a lot of memory, takes up CPU time and the calculation of an average value counts in all values. Therefore, it can be advantageous to use a digital filter. See more in chapter 13.4 Digital Low-Pass filter, page 100.

10 Split-up in program modules

A division/Split-up into program modules and functions is one of the basic and important building blocks in a structured PLC program. They individually consist of a small piece of PLC code, to be used whenever needed. The program modules need individually to have an indicative name as explained in chapter 6, page 29.

In order to have a good structure and a structured program, it is a good rule of thumb only to have max. 20 – 25 of lines PLC code in each program module. It is much easier to work with small pieces of PLC code rather than one big giant program.

Furthermore, it is easier to move and correct the program. The chronological sequence might have to be changed or some program modules must be made inactive when fault and error searching (insert // before the program module name).

Below a main program is shown, where three sub-programs are used:

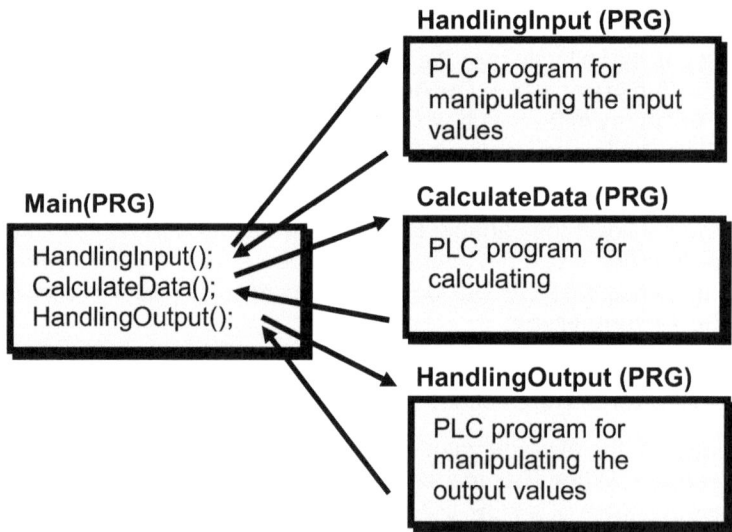

The main program **Main** is executed once in each program-scan. In the **Main** the program module **HandlingInput** is applied, as it is positioned first and when the whole PLC code in this program module is executed, **CalculateData** comes up next. Finally, the **HandlingOutput** program module is executed.

When the program-scan time ends, then it must be repeated. If the program-scan time is 50 [ms], the Main program is executed every time 50 [ms] has passed. It is important that the total time of execution for all four program modules is below 50 [ms]. If the program modules contain large arrays or many calculations, all program modules might not be executed within the scanning time. In order to solve that, the program-scan time is increased or the programs are to be examined to see whether some parts can be reduced or redesigned. It must to be remembered that program modules have a variable time of execution as IF-statements or CASE statements can include conditions, which are not always fulfilled. Therefore, the worst case scenario of program execution time must be taken into consideration and plan extra time for each program scanning.

There are many ways of splitting up a program. Below an inspiration list is made:

- Sensors on one side of the machine
- Digital input from electrical contacts, switches and breakers
- All motors for ventilation
- Handling values from HMI (HMI = user control panel)
- Procedure of program commissioning
- Execution of program sequences
- Program module for stopping the machine
- Alarm surveillance / Alarm supervision
- Fault and error handling on the machine
- Handling data communication to other PLCs

10.1   Functions

Functions are important building blocks in a PLC program. A function contains a small PLC program code, used again and again.

A function 'call' to **MyFunction** is carried out as follows:

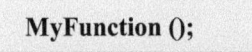

```
MyFunction ();
```

The above function 'call' does not contain parameters, because the brackets are empty. It is possible to make use of a function consisting of one or more input values (parameters), on which the function consequently has to handle or use for calculating. When the function 'call' is ended, one or more values (parameters), which the overall program can work on, must be returned by the function

Below is shown a function 'call' including two parameters (input values), 12 and 3:

MyFunction1 (12, 3);

The advantage by applying functions is that the PLC code can be reused. PLC code reuse reduces the size of the program, creates fewer syntax faults and is easier to work with for others.

It is possible to make calculations before a function 'call'. Below two numbers (3+7) are added just before the function is 'called':

MyFunction2 (3 + 7);

The calculation is carried out before the function 'call' and the input value to the function is therefore 10.

If the function is to return values when the function is ended (a result of one or more calculations), it can only be carried out by using variables, as the function have a 'shelf' in the memory to deliver the value. When a function is 'called' with an input variable, the function will collect the value in the variable 'shelf' in the memory and deliver a copy of the variable into the function.

Below is shown a program 'call' to a function with variables:

> **MyFunctionInOut (Var1:= ValueIn, Var2=>ValueOut, Var3:= ValueInOut);**

The three variables **Var1**, **Var2**, **Var3** are created inside the function and they are created with the following variable scope:

Variabel	Scope	Assigment
Var1	IN	:=
Var2	OUT	=>
Var3	IN_OUT	:=

ValueIn is a value going into the function and it is written as follows:

> **Var1 := ValueIn;**

The value which must out of the function must be delivered in the variable **ValueOut** and it is written as follows:

> **Var2 =>ValueOut;**

The variable which is both going in and out of the function delivers the address pointing at the 'shelf' in the memory. Must be written as follows:

> **Var3 := ValueInOut;**

Notice how the assignment signs "=>" and ":=" are used at function 'calls'.

How to 'call' to an ARRAY with functions:

A program call to function no. 4 in an ARRAY of functions is carried out as follows:

> **MyFunction [4] (ValueIn);**

10.2 Function (FC) and Function Block (FB)

There are two function types in a PLC:

Function (FC)

Function block (FB)

Function (FC) is a PLC code excluding static data, meaning that all local variables lose their value when the function is ended. The variables are initialized again the next time the function is 'called'. The function typically carries out a mathematical calculation and returns the finally calculated value.

Function block (FB) is a PLC code including static data. The local variables retain their values between each 'call' to the function. An example could be a function used as an hour counter (number of operation hours, also called TACHO HOURS) on a motor and therefore the local variables must retain their values after having ended the function. The function could also count the number of motor starts per hour or time for the next service visit on a motor.

Format for **FUNCTION**

```
FUNCTION <Name> : <RetDataType>
   VAR_INPUT
      <Variables>
   END_VAR
   VAR_OUTPUT
      <Variables>
   END_VAR
   VAR_IN_OUT
      <Variables>
   END_VAR
   VAR
      <Variables>
   END_VAR
      <Implementation> //Write code here
      <Name> := 123;  //set return value
END_FUNCTION
```

Format for **FUNCTION_BLOCK**

```
FUNCTION_BLOCK <Name>
   VAR_INPUT
      <Variables>
   END_VAR
   VAR_OUTPUT
      <Variables>
   END_VAR
   VAR_IN_OUT
      <Variables>
   END_VAR
   VAR
      <Variables>
   END_VAR
      <Implementation> //code here
END_FUNCTION
```

Implementation of a function starts by the keyword: FUNCTION and a function block is starts by keyword: FUNCTION_BLOCK. Afterwards, in the frame **<Name>**, the name of the function is written. It must be an indicative name (see chapter 6, page 29 about 'naming') related to what the function must carry out. The returning data type is written in the **<RetDataType> section** field, as the name of the function works as the returned value.

Notice that the **<RetDataType>** field cannot be used in FUNCTION_BLOCK.

The sections with VAR_INPUT, VAR_OUT and VAR_IN_OUT must contain variables that goes in and out from the function. When VAR_INPUT is used, the function is copying the variable and is working on it inside the function – without destroying the original variable. If VAR_IN_OUT is used, the address of the variable is delivered and the function is working directly on it inside the function, and must be used carefully.

If a function needs to work with STRUCT or ARRAY the VAR_IN_OUT must be used.

The sequence of variables listed in the function indicates the sequence of the variables when 'calling' the function.

The section VAR contains the local variables, only to be used internally in the function. When a function 'call' is carried out, the local variables are created every time the function is 'called' and laid down again, when the function is ended.
Remember that variables must be initialized (be set at a commissioning value, e.g. 0) to insure which value the variables have, when the function is 'called'.
If the function is to save local variables at each 'call', a FUNCTION_BLOCK must be used or VAR_IN_OUT is to be used, so that the function works on variables, created outside the function.

The PLC code, the function is to execute, is written in the **<Implementation>** section.

When a FUNCTION is used, the return parameter is to be set *BEFORE* the function is ending. It is set by using the name of the function, assigned the return parameter. In the above format, the return parameter is set to 123. Only one return parameter can be set in this way. If more return parameters are needed, the VAR_OUT and/or VAR_IN_OUT are to be used.

In order to make the PLC program clear and readable, it is recommended that a FUNCTION and FUNCTION_BLOCK only contain so much PLC code, than seen on the screen during programming, i.e. up to 20-25 lines. If the PLC code is longer, there is a need of creating yet another function.

A function ought only to provide max. 8 parameters (variables) in the function 'call' as it can be difficult to overview more. If more variables are still needed, a **STRUCT** where many variables are collected can be created and sent together to the function. If **STRUCT** is used, the variable must be positioned in the **VAR_IN_OUT** section.

A function is to be seen as a black box. When the function is really working, everything in the box goes on automatically.

IMPORTANT: A function must never make a function 'call' to the function itself!

There are many possible functions. Below is a list of inspiration:

* Conversion between measuring units
* Counting hours on motors
* Calculation of expected time for service
* Calculation of speed on conveyer belts
* Scaling analogue values
* Volume calculation of tanks
* Searching for min and max values in an array
* Mathematical calculations
* Pulse generator
* PID regulator
* Alarm surveillance/ supervision on mechanical components
* Conversion of values from temperature sensors
* PLC code to be reused in other PLC program
* Calculation of optimum operation range for frequency converters
* Estimation of expected time of production

The difference between functions and program modules is the fact that functions often make calculations or data handling on single components, whereas program modules split up the entire program. The program modules use relevant functions and function blocks to solve concrete tasks.

Typically, it is easier to reuse functions rather than program modules.

The following pages contain examples of functions.

10.3   EXAMPLE: FC to conversion of temperature

This example is showing an implementation of a function, converting temperatures from Celsius (Centigrade) units to Fahrenheit units. It is implemented in a function as it is a mathematical calculation being reused many times in the program.
A FUNCTION, named **fcTemperatureCalculateCtoF**, is created which returns a REAL variable, as calculations - including temperature - are often decimal numbers. The name of the function starts with 'fc' to show that it is a function. The rest of the name is chosen, because it fits well in with what the function is able to do. The naming begins with a noun: **Temperature** and a verb: **Calculate** and the letters: **C to F** showing that a conversion takes place: Celsius (Centigrade) temperature is converted into Fahrenheit temperature.

```
FUNCTION fcTemperatureCalculateCtoF : REAL
VAR_INPUT
   TemperatureC: REAL;
END_VAR
```

The function provide one single input parameter named **TemperatureC** and it is created in the VAR_INPUT section as shown below and is of the data type REAL, because the parameter (Celsius/Centigrade temperature) is a decimal number.

The PLC code inside the function is shown below:

```
//This function convert a Celsius temperature to a Fahrenheit temperature
//Input parameter (REAL) is in Celsius
//Out parameter (REAL) is in Fahrenheit
fcTemperatureCalculateCtoF:= (TemperatureC * 9/5) + 32;
```

The formula used for the conversion is found on the Internet.

As can be seen, the return parameter is the name of the function with the data type REAL. Comments in the beginning of the function are made to explain other users what the function is able to do; Good programming always includes writing comments in the beginning of the function, even if the function name is self-explaining.

Below is shown how the function can be used, where **TemF** is a **REAL** datatype:

```
TempF := fcTemperatureCalculateCtoF(23.6);
//The value is copied to TempF and is 74.48 (REAL data type)
```

The function is 'called' with the value of 23.6 (Celsius/Centigrade temperature of degrees C). The function returns the calculated Fahrenheit value in the variable **TempF**.

In order to test the function, a corresponding calculation website is found on the Internet and large and small values are to be inserted in order to test whether the function is working as expected. It is always important to test the function thoroughly, as it can be difficult to find faults, when the program becomes larger and made up by many functions and program modules.

10.4   EXAMPLE: FC to calculate average

The following chapter shows an example of a function, which calculates the average pressure on two sensors. There is no need of saving values and therefore it is a FUNCTION which is created. The function is named **ValueAverage** with two input parameters: **Value1** and **Value2**; both of the data type REAL. Even if it is an average of two pressure measurements, one single function is created with one general name, so that the function can be reused - with no confusion of the name of the function. The calculated value is of the data type REAL and this is defined as a return parameter for the function by writing REAL in the first line of the code as shown:

```
FUNCTION ValueAverage : REAL //REAL is return parameter data type
VAR_INPUT
  Value1, Value2 : REAL; //Input parameters to the function
END_VAR
VAR
  Sum : REAL;  //Local variable for temporary calculation
END_VAR

Sum := Value1 + Value2;  //Total sum
Sum := Sum/2; //Average
ValueAverage := Sum;  //Set the return parameter
```

In the last line the return value is set. It means that the **ValueAverage** has a value before the function is ended, which makes sure that the calculated value can be used outside the function. The variable **sum** is a local variable and can therefore not be used outside of the function. This creates a good program structure and is a 'black box' for an average calculation of two values.

The shown variables below are as follows: **Avg1**, **Avg2**, **Avg3**, **Sensor1Pressure**, **Sensor2Pressure** are all in the data type REAL, as it is the data type which the **ValueAverage** function uses as input parameters and return parameters.

Examples of how to use the **ValueAverage** function:

```
//Example #1: Use the functions variable names
Avg1 := ValueAverage(Value1 := 85.1, Value2 := 17.6);
//Example #2: Use value only
Avg2 := ValueAverage(85.1, 17.6);

//Assign value to main variables
Sensor1Pressure := 85.1;
Sensor2Pressure := 17.6.;

//Example #3: Use main variables
Avg3 := ValueAverage(Sensor1Pressure, Sensor2Pressure);
//Example #4: Combination of #2 and #3
Avg4 := ValueAverage(Value1 := Sensor1Pressure, Value2 := Sensor2Pressure);
```

When a FUNCTION is used, all input parameters must have a value. If a FUNCTION_BLOCK is used is not necessary that all input parameters must have a value. However, it is a good idea to asign all input parameters a value, because it indicates that the programmer decides on which parameters to use and remember them.

The parameter sequence for the function 'call' is important and it is the same, as when the function was created. Therefore, **Value1** has to be positioned first and then **Value2** afterwards.

The PLC code in a FUNCTION and FUNCTION_BLOCK *MUST* take into consideration that input parameters are missing, or parameters are positioned outside the permitted range, or are invalid. The PLC code inside the function must be stable and be able to be executed, even if input parameters are missing, are wrong or invalid.

Vice versa, the programmer, who implements the function, has also an obligation to make sure that FUNCTION or FUNCTION_BLOCK will be 'called' with valid and legal input parameters. Further, the programmer must write a description of the function explaining the mode of operation and input parameters. Finally, the function is not finalized until it is tested!

11 Working with text and chars, STRING

STRING is the data type to be used, when working on texts. Below is shown some areas, where a PLC must be working on texts:

Reading dynamic texts and digits on HMI/SCADA (user operation):

- Online changes of languages on user operation panels (e.g. change between Danish and English user interfaces with no changes in the PLC code (Multi language change)
- Messages and instructions to the user: production information, typing passwords, reading of letters, time/date, alarm texts

Handling files and database data:

- Reading data from files on a hard disk (e.g. settings of equipment and instrumentations, configuration files, set points)
- Data logging of measuring data or events (e.g. changing settings or mechanical condition changes)
- Language texts to be read from hard disk or flash card
- Messages to/from production systems (ERP, SAP, MES)
- File names, folder names, e-mail

Data communication between PLC/PC/Instruments

- Instruments send data in ASCII (e.g. BAR/QR codes, RFID, TAGS)
- Information to label printer (e.g. labels to boxes, production dates)
- SMS (alarms/command to/from mobile phones/smartphones)
- Numbers with many digits and mixed with letters
- Measuring data, alarms, information from automation equipment

The following data types exist:

Data type	Description
CHAR	Contains one character only (ASCII) (8 bit)
WCHAR	Contains a wide character (16 bit) (UNICODE, ISO 10646)
STRING	ARRAY of CHARS [0..254], for sentences (254 is max.)
WSTRING	ARRAY of WCHAR [0..254], for sentences (254 is max.) Used for PLC controls handling multi languages on HMI (**H**uman **M**achine **I**nterface) (UNICODE, ISO 10646)

NOTICE:

Use only STRING when necessary, as it requires PLC power and uses a lot of memory.

Only create STRING of the length needed.

Not all PLC types provide the data types CHAR and WCHAR. If there is a need of a variable with one single sign (character) after all, create one STRING[1] or one BYTE.

IMPORTANT: The length of a STRING is defined by counting characters until 0 (zero) is found in the ARRAY (Some programming languages place the length of the STRING at positon zero, which is important if a PLC communicates with another equipment)

A STRING shows characters by using an ASCII table. Integers are saved in ARRAY, as a CPU is only able to save data in integers. Below is shown an ARRAY with integers and the corresponding characters from the ASCII table:

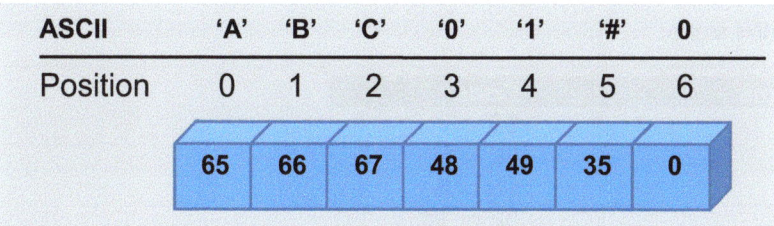

A PLC provides typically a maximum length of 255 characters in a STRING. If a text is longer than 255 characters, the text can be split up in more STRINGSs.

A STRING can be created with or without a fixed length as shown below:

```
PROGRAM DemoString
VAR
     szDemo: STRING          := 'Having no fix length'
     szDemoFix: STRING[35] := 'Fixed length string';

     szEmpty: STRING         := '';      //String without text
     szDemoW: WSTRING     := "This is a UNICODE string";
END_VAR
```

If NO length is indicated – as is the case when using **szDemo** – then the PLC uses 254 bytes in the memory + 1 (Zero sign for ending STRING is included).

If a fixed length is set – as is the case when using **szDemoFix** – then the PLC uses the fixed length – here 35, to the memory + 1 (Zero for ending STRING is included).

The above statement means that the best choice is to set a max length on all STRING. As texts, however, can be dynamic during the execution of a program, chal-lenges can occur using a fixed STRING length. It can e.g. be the case when making language changes online, where texts can be 50 % longer, when changing from a English text to French text.

It is not possible to write a text with double citation sign: *A "big" test*. A control sign must be written ($ sign) before the text: A *$"big"* test.

Table with control signs:

Control signs	Code
Dollar sign	$$
Line shift	$L or $l
New line	$N or $n
New page	$P or $p
<RETURN>	$R or $r
<TAB>	$T or $t
Citation sign	$'
Double citation sign	$"

11.1 EXAMPLE: FC with STRING

Below is shown an example of how a FUNCTION with STRING is defined:

```
FUNCTION StringDemoFUN : STRING
VAR
      str4: STRING;
END_VAR
VAR_INPUT
      Str1: STRING;
END_VAR
VAR_OUTPUT
      str2: STRING;
END_VAR
VAR_IN_OUT
      str3: STRING;
END_VAR
```

```
str2:= 'STR 2 string';
str3:= 'STR 3 string';
str4:= 'STR 4 string';
StringDemoFUN:= Str1; //Set return parameter
```

Program 'call' to the function **StringDemoFUN:**

```
MainStr:= 'Hello World';
Mstr1:= StringDemoFUN (str1:=MainStr, Str2=>MStr2, str3:=Mstr3);

//Contents of the variables are:

//MStr1 = 'Hello World'.
//MStr2 = 'STR 2'   //Because STRING length is 5: Mstr2[5]
//MStr3 = 'STR 3 string'.
```

NOTICE:
The variables **Mstr1**, **Mstr2** and **Mstr3** are all created as a STRING data type. If a variable is created with a fixed length of e.g. 5, as **Mstr2**, it will only contain 5 characters, even if the string str2, created inside the function contains 12 characters.

11.2 Standard functions, STRING

The built-in standard STRING functions are shown below. Some PLC types provides more functions and they can be found in the manufacturer's programming manual.
If a certain STRING function is needed, the programmer must often implement it himself or try to find it on the internet.

The max length for STRING in the standard functions is 255 characters.

CONCAT

Connects two STRING
STR2 is inserted after **STR1**

Str3:= CONCAT (**STR1** := 'AB', **STR2**:='CD');
//Str3 = 'ABCD'   Str3:= CONCAT ('AB', 'CD');

INSERT

Inserts a STRING in another STRING at a certain position. **STR2** is inserted in **STR1** at **POS** position

Str3:= INSERT (**STR1**:='ABCD', **STR2**:='EFGH', **POS**:=2);
//Str3 = 'ABEFGHCD'   Str3:= INSERT ('ABCD', 'FEGH', 2)

DELETE

Delete some part(s) of a STRING. **IN1** is the STRING
From position **POS** the amount, which **LEN** indicates, is deleted

Str3:= DELETE (**IN1**:='ABCDEFG', **LEN**:=2, **POS**:=3);
//Str3 = 'ABEFG'   Str3:= DELETE ('ABCDEFG', 2, 3);

REPLACE

Replaces some parts(s) of a STRING. The amount of **L** characters in **STR1** is deleted. **STR2** is inserted from position **P**

Str4:= REPLACE (**STR1**:='ABCDEFG', **STR2**:='X', **L**:=2, **P**:=3);
//Str4 = 'ABXEFG'   Str4:= REPLACE ('ABCDEFG', 'X', 2, 3);

FIND

Find a STRING in another STRING
STR2 in looked for in **STR1**. An INT is returned with a position in which **STR1** first was found. If nothing is found, 0 (zero) is returned. The FIND function is dependent on big letters (upper case) and small letters (lower case)

Int1:= FIND (**STR1**:='ABCBCDEFG', **STR2**:='BC')
//Int1 = 2   'BC' is found first at position 2

LEN

LEN finds the length of a STRING. Counting numbers of characters in **STR**
An INT with the length is returned

Int2:= LEN (**STR**:= 'Demo') ;
//Int2 = 4 or this can be used: Int2:= LEN ('Demo') ;

LEFT

LEFT keeps some part(s) of a STRING from left. The first parameter **STR** is STRING and the second parameter **SIZE** is the amount of characters, which retained.

Str6:= LEFT(**STR**:='1234567', **SIZE**:=2);
//Str6 = '12' or this can be used: Str6:= LEFT('1234567', 2);

RIGHT

RIGHT keeps some part(s) of a STRING from right. The first parameter **STR** is STRING and the second parameter **SIZE** is the amount of characters which retained.

Str7:= RIGHT (**STR**:='1234567', **SIZE**:=2);
//Str7 = '67' or this can be used: Str7:= RIGHT('1234567', 2);

MID

MID keeps some part(s) of a STRING. The first parameter **STR** is STRING, **LEN** is the length of what is retained and **POS** is the start position of what is retained.

Str8:= MID (**STR**:='1234567', **LEN**:=2, **POS**:=3);
//Str8 = '34' or this can be used: Str8:= MID('1234567', 2, 3);

NOTICE: In not the all PLC types, relational operators (See chapter 7.2, page 39) can be used directly on STRING in IF statement, as STRING is an ARRAY. The built-in FIND and LEN functions must be used when comparing texts:

```
Str1 := 'abc';
Str2 := 'abc';

IF Str1 = Str2 THEN
   Str3:= 'Same';
END_IF;
```

```
Str1 := 'abc';
Str2 := 'abc';

IF FIND (Str1, Str2) > 0 THEN
   IF LEN (Str1) = LEN (Str2) THEN
      Str3:= 'Same';
    END_IF;
END_IF;
```

OK

For converting numbers, the built-in data type conversion functions can also be used for STRINGS (see chapter 8.6, page 49) as seen below:

```
myint:= STRING_TO_INT('123');
myreal := STRING_TO_REAL ('12.45');
myStr1 := REAL_TO_STRING (23.67);
```

Before conversion functions are 'called', the string (which is the input parameter) must be controlled, so that the function does not receive characters in a string which is not convertible. It might be unclear what is going to happen if the PLC program must convert e.g. 'ABC' to a REAL data type. Functions exist, which can be called **IsNumber** on the internet, being able to be used to control whether the contents of a string is a number.

> **IMPORTANT:** In some PLC types, STRING standard functions are not 'thread safe'. This means that the best choice is to only make use of them in a PLC code, being executed in the same task.

As STRING is an ARRAY, it is possible to insert a few characters directly. Below three different examples are shown, because the different PLC types handle this differently:

```
str1:= 'My String';       //The beginning string
str1[2]:= 'A';            //Example 1, insert 'A' into position 2 in str1
str1[2]:= 65;            //Example 2, insert, where 65 is 'A' in the ACSII tabel
str1[2]:= F_toASC('A'); //Example 3, use a build-in function named F_toASC

//The resulting string is 'MyAString' where 'A' is overwriting <SPACE> in str1
```

12 Built-in standard functions

This chapter describes a range of built-in standard functions. When they are to be used, depends on which tasks must be solved and it must be noticed that the functions can be named differently in the different PLC types, found on the market.
If the standard built-in functions are used it can be more difficult to copy the PLC code to other PLC types, as the code might be adjusted.

12.1 One time program execution: First ScanBit

A need might occur that some part(s) of the PLC code must be executed only once and only just right after powering up the PLC (PLC turn on). It could be digital outputs that must be initialized at a certain value in order to make sure that e.g. signal lamps turn on right or a valve is set to OFF. Maybe internal variables, counters and arrays must be reset to zero.

Some PLC types provides a first-scan-bit or *FirstCycleBit* for this purpose. However, if the PLC does not provides such feature, the below PLC code can be used:

```
VAR
    FirstScanBit : BOOL := FALSE; //#2
END_VAR

//Set first scan bit
IF FirstScanBit = FALSE THEN //#3
    // First code, initialization code here or call to a program module
    // code here will be executed only once #1
    FirstScanBit:= TRUE; //#1
END_IF;
```

Mode of operation:

A BOOL variable **FirstScanBit** is created which is initialized in the variable section to FALSE (see **#2**). This causes that first-scan-bit is *always* FALSE, when starting up the PLC. When the PLC code is executed the first time, the PLC code within the **IF** statement will be executed as **FirstScanBit** is FALSE (see **#3**). When **FirstScanBit** is set to TRUE the PLC code added by **#1** is not executed again.

12.2 Edge detection (One shot): R_TRIG, F_TRIG

There is often a need for a PLC code only to be executed once in a certain action. It can be a sensor contact which is activated and a belonging PLC code then needs to be activated (A sensor e.g. counting objects on a conveyer belt). When the sensor contact is activated the code will be executed several times due to the mode of operation on which a PLC executes a program, unless take care of that.

This can be solved by a function having many names: Oneshot, edge detect, OSRI.

Two standard function blocks exist, making sure that a code is only executed once:

R_TRIG is used for detecting a rising edge, positive flank (signal: 0 **=>** 1).
F_TRIG is used for detecting a falling edge, negative flank (signal: 1 **=>** 0).

The function blocks provide an input parameter **CLK** and an output parameter **Q**, both of the data type BOOL.

Below an example is shown (**EXAMPLE 1**):

```
PROGRAM MAIN
VAR
  B1OneShot : R_TRIG;      //One shot for the B1 sensor input
  B1 : BOOL;               // B1 is the sensor input
END_VAR

//EXAMPLE 1: One shot is using an instance of R_TRIG (Positive flank)
B1OneShot (CLK := B1); //'Call' the function block

IF B1OneShot.Q = TRUE THEN
  // Run the one shot PLC code here #1    .
  // A program module or a function can be called here
END_IF;
```

The mode of operation is as follows:

B1 becomes **TRUE** when the sensor input is activated, and **B1** is the input parameter to the **B1OneShot** function. It causes that the BOOL variable **B1OneShot.Q** is **TRUE** in the program-scan, where **B1** became **TRUE**.
In the following program-scan, **B1OneShot.Q** is automatically set to **FALSE** by the built-in R_TRIG function. The PLC code in #1 section is therefore only executed once.

EXAMPLE 2 is a do-it-yourself solution without using R_TRIG. Here the physical contact is the input **B1** and when it is 1 (activated by e.g. a contact or a sensor counting objects on a conveyer belt) at the same time as **B1Old** is 0, the PLC code will be executed, marked by **#1**. When the code in **#1** is executed, **B1Old** is set to 1. In the next program scan the code is not executed. When **B1** again is 0, **B1Old** is set to 0.

Below the example is shown:

```
PROGRAM MAIN
VAR
  B1: BOOL;
  B1Old: BOOL;
END_VAR

//EXAMPLE 2: Using own created PLC code
//Detect on rising edge
IF  B1 = 1 AND B1Old = 0 THEN
  B1Old:= 1;
  //Insert PLC code here to run only once #1
END_IF

//Reset edge detection
IF B1 = 0 THEN
  B1Old:= 0;
END_IF;
```

It is easier to copy **EXAMPLE 2** than **EXAMPLE 1** to another PLC as the different PLC types have different one shot standard function blocks.
The execution for **EXAMPLE 2** can be illustrated on the below time diagram:

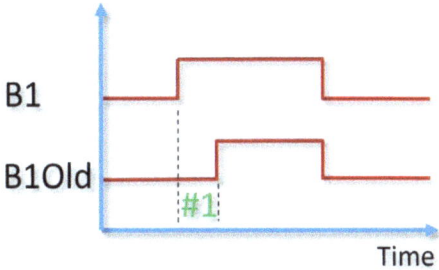

The PLC code **#1** is executed immediately after a rising edge on **B1**.

12.3 Counting functions: CTU, CTD, CTUD

A PLC provides three built-in counting function blocks:

CTU, can count upwards
CTD, can count downwards
CTUD, can count both upwards and downwards

Below is shown how CTU functions block is used in ST- programming:

```
PROGRAM MAIN
VAR
    myCTU        : CTU;    // Counter UP function
    S1           : BOOL;  // Activate count
    K1           : BOOL;  // TRUE when count finish
    i            : WORD; // Only for demo and test
END_VAR
```

```
// Example 1, counter using the CTU function block
myCTU (CU:= S1, PV:= 12, RESET:= myCTU.Q); //Counting to 12, auto reset

IF myCTU.Q THEN //Counter done?
  K1 := TRUE; //#1
END_IF;

i:= myCountDemo.CV; //Readout current count value
```

A variable **myCTU** with the data type CTU is created, which is a built-in standard function block, being able to count upwards. **CTU** has three input parameters: **CU** (counting), **RESET** (resetting counter to 0, on the positive flank) and **PV** (max. counting value, where 0 is included in the counting) and two output parameters **Q** (max. counting quantity achieved) and **CV** (present counting value). **CU** is set to a BOOL value with **S1**, which can be a physical contact and every time it is activated, counting is set 1 up. When the counter has counted to 12 (counted from 0 to 11) **Q** is TRUE and an IF statement sets **K1** to TRUE. **K1** can be a lamp (see #1).

In order to make the counter reset automatically, when the max value is reached and restarts, **RESET:= myCTU.Q** is inserted in the parameters for the **myCTU** function.

The advantage of the CTU function block is the fact that it provides a R_TRIG built-in in the **CU** input. The disadvantage is the fact that it counts internally on a WORD variable and can therefore only count up to 65535. If CTU is used for counting objects on a machine, producing an object per minute, an overrun occurs on the internal counter after some days calculated as follows:

60 [parts/hour] => 1440 [parts/day] => 65535/1440 => 45.5 days.

Below is a solution being able to count on a DWORD (double WORD) variable:

```
PROGRAM MAIN
VAR
    S1_trig: R_TRIG;     // One short
    S1: BOOL;            // Activate count
    K1: BOOL;            // TRUE when count finish
    i: DWORD := 0;       // Counter
END_VAR
```

```
// Example 2, Counter with DWORD
S1_trig (CLK:= S1);  // Calling R_TRIG

IF S1_trig.Q THEN   //Count up if positive trig signal
  i:= i + 1;
END_IF;

IF i >= 12 THEN //Counter done? #1)
  K1:= TRUE;   // Set output
  i:= 0;           // Reset counter
END_IF;
```

K1 becomes **TRUE** when the counter has counted up to 12 and at the same time the counter variable **i** is set to 0.

Remark: #1) In order to create more stable PLC code use "**>=**" instead of only "**=**".

The mode of operation for the two examples (Example 1 vs Example 2) is the same, Example 2 is, however, more usable:

- It can count up to 4.29 billion
- It is independent of the type of PLC

A counter can be used for counting produced parts, amounts of starts on a pump, amount of pulses from instrumentations: e.g. energy meter or a flowmeter.

12.4 Repeated program 'call' and timer delay: TON, TOF

In a PLC program, some equipment must run in for certain time. A motor must e.g. run for 30 minutes per hour, the light in a staircase must switch off automatically after a certain time or a stop watch must be controlled. It could be an alarm signal from a level sensor in a tank, not appearing until after a certain period, as wave motions in the tank can affect the level sensor measuring if an error may occur. A timer solves these problems:

Two types of standard timers exist in a PLC:

TON (On-delay timer, TOD-TimerOnDelay, ON delay)	Delayed connection
A **TON** timer function block sets a **BOOL** variable **Q** to **TRUE** after a certain period of time indicated by **PT**. Can be used if a component must have a signal at a certain period of time in order to start. Used for Noise Attenuation in an ON/OFF contact. The time where **IN** is activated must be longer than **PT**.	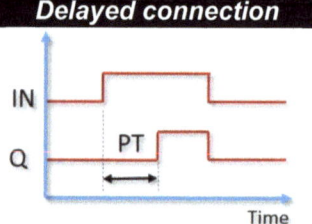

TOF (Off-delay timer, OD Off-Delay, OFF delay)	Delayed defection
A **TOF** timer function block sets a **BOOL** variable **Q** to **FALSE** after a certain period of time indicated by **PT**. Can be used for light in a staircase or toilet ventilation, where the system must be switched on for a period of time after a contact **IN** has been activated.	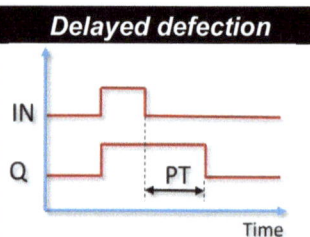

A timer is a built-in functions block, provides two input parameters (**IN** and **PT**) and two output parameters (**Q** and **ET**). The positive flank on **IN** starts the timer and on **PT** the time period is indicated. **Q** is the signal output and **ET** shows the current time.

Below a timer is shown, being active for 100 milliseconds after **S1** has become **FALSE**.

```
VAR
    S1TimerTOF : TOF;    //Create timer
    S1 : BOOL; //Switch
END_VAR

S1TimerTOF (IN:= S1, PT:= T#100ms);
IF S1TimerTOF.Q = TRUE THEN
    //Code here will be active in 100 [ms] after S1 = FALSE
END_IF;
```

Example 2, shown below, indicates how a timer can be implemented including an automatic restart. The timer is active for 10 seconds and restarts automatically.

```
PROGRAM MAIN
VAR
    MyTimer:       TON;              //Create timer
    TimerCurrent: TIME;             //Only used for readout
END_VAR

//Example 2, timer automatic restart
Mytimer(IN:= NOT Mytimer.Q, PT := T#10S);  //Start or restart timer.

IF Mytimer.Q = TRUE THEN
    //Write code here to be called each 10 sec
END_IF;

TimerCurrent := MyTimer.ET; //Only for readout
```

The mode of operation is as follows:

Two variables are created:

MyTimer: the data type is a TON function block, which must be used, because the timer always has to remember how long time it has been active.
TimerCurrent: only used to be able to read the current value of the timer – an efficient tool to make it all work.

The current value on the timer is to be read on the last line in the PLC code and is read out by copying the value, indicated on **MyTimer.ET**, to **TimerCurrent**, which is created with the data type TIME, because it is of the same data type as **MyTimer.ET**.

When the timer is active, **MyTimer.Q** = FALSE and if the timer has expired, **MyTimer.Q** = TRUE and the timer stops. The timer restarts automatically, because **IN** is the inverted value of **MyTimer.Q** (use NOT before **MyTimer.Q**).
The parameter **PT** setup the time delay, where time is indicated by **T#** and a digit (here 10) followed by the SI-unit (s = second, ms = millisecond, h = hour).

Timer as a task
Another possibility of implementing a timer of 10 seconds is to create a PLC-task, which is e.g. 'called' every 1 second. In the program module, which the task is 'calling', a counter is created and set to zero when 10 is reached. If the PLC scan-time is 10 [ms] count to (10 [s] / 0.010 [task/s]) = 1000, before 10 seconds have passed.

13 Special functions and structures

This chapter describes a range of special functions and often used structures.

13.1 Simple queue structure

This example describes the simplest implementation of a queue. A queue is used when e.g. there are many packages on a conveyer belt, waiting for treatment by a machine in a large plant. The packages often require information like e.g. weight, receiver, size or content. A weight gives information about a package and the information must be saved in a queue so that the information can follow the package through the plant. If the package has a readable bar code, it is not necessary to implement a queue, as the information about the package can be given from a common data base - e.g. the company's production control system, often named **M**anufacturing **E**xecution **S**ystems (**MES**) or **M**anufacturing **I**nformation **S**ystems (**MIS**)

When implementing a queue it is required that the objects do not change sequence in the queue. If the packages provides e.g. a bar code or any kind of ID the packages may change sequence in the queue.

An **ARRAY** must be used and be created with the max length which is predicted for the queue. **ARRAY** must not be created too long, as it uses a lot of memory and takes longer time to execute for the PLC

In order to make it simple in the flowing example, an **ARRAY** with 6 positions of the data type **INT** is created (see figure below). Firstly, all the **ARRAY** positions is initialized to -1, as -1 can be used for checking whether the position is empty:

```
PROGRAM MAIN
VAR
        Que: ARRAY[QueMin..QueMax] OF INT;
        n: INT; //Counter to FOR loop
END_VAR
VAR CONSTANT
        QueMax: INT := 5;
        QueMin: INT := 0;
END_VAR

FOR n:= QueMin TO QueMax DO
   Que[n]:= -1; //Init ARRAY
END_FOR;
```

The number above the **ARRAY** shows the no. of position:

0	1	2	3	4	5
-1	-1	-1	-1	-1	-1

The **ARRAY** is now filled with three values (23, 35, 71). **ARRAY** is filled from left to right, so that the value inserted first (here 23), is positioned all the way to the left (here 71) and the value inserted the latest, is positioned all the way to the right on position 2 as shown below:

0	1	2	3	4	5
23	35	71	-1	-1	-1

Inserting values in the queue can be carried out with this PLC code, where the **ARRAY** is named **Que:**

```
Que [0] := 23;
Que [1] := 35;
Que [2] := 71;
```

The oldest value in the queue is 23 and is also the value which is taken out first. In order to keep control of the queue, the simplest way to do so is to make sure that the oldest value is always positioned at position 0.

When the oldest value is taken out, all the values are moved one position to the left. The next value which is to be taken out is therefore now 35.

When the oldest value is moved from the queue, all numbers must change their position in order to make sure that the oldest value is now positioned on position 0 (zero).

A **FOR** loop is used for moving all the values to the left. Values must always be moved to the left in order not to overwriting the values which already exist in the queue. The **FOR** loop must be executed one time less than the max amount in **ARRAY**, as shown in the example below:

```
FOR n:= 0 TO 5 - 1 DO
    Que [n]:= Que [n + 1];
END_FOR;
```

As can be seen, **ARRAY** provides 6 positions and is copied 5 times and the **FOR** loop is therefore executed 5 times and can be illustrated as follows:

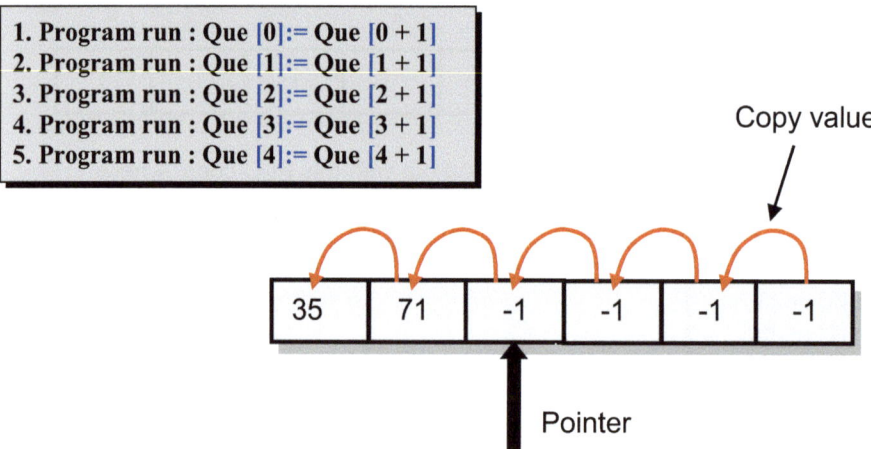

```
1. Program run : Que [0]:= Que [0 + 1]
2. Program run : Que [1]:= Que [1 + 1]
3. Program run : Que [2]:= Que [2 + 1]
4. Program run : Que [3]:= Que [3 + 1]
5. Program run : Que [4]:= Que [4 + 1]
```

Copy value

| 35 | 71 | -1 | -1 | -1 | -1 |

Pointer

In order to keep control on which position the next value must be inserted at, a variable must be used, called index or pointer. It starts by pointing at position 0, because the queue is empty. Every time a new value is inserted into the queue, the pointer is moved one position to the right, and if a value is removed (taken out) from the queue the pointer is moved one position left.

The disadvantage of this simple queue is that a lot of time is used on executing the whole queue to 'push' the values ahead every time a value is taken out. To solve the problem a circular buffer is used or a ring buffer, using pointers instead of moving data, every time a value is taken out or inserted.

A queue is often called **FIFO** – meaning **F**irst **I**n **F**irst **O**ut. The value which is inserted as the first value has to get out first. This is described in the next chapter.

13.2 FIF0 – First In First Out

The former chapter described the implementation of a simple queue, where all positions moved every time a value was moved from the queue. This chapter describes a queue, where the values ARE NOT MOVED, when a value is taken out. This makes the PLC code effective.

An effective **FIFO** consists of an array and two pointers, pointing on a position in the array, as shown the below illustration:

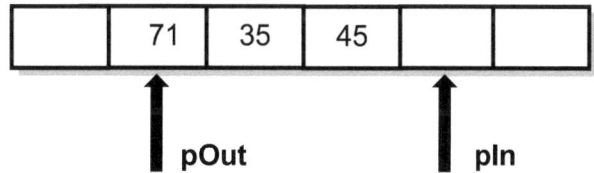

The pointer **pOut** is pointing at the value which must come out of the queue first and the pointer **pIn** is pointing at the next free position in the queue. Every time a value is removed from the queue, the pointer **pOut** is moved one position to the right. Every time a new value is inserted, where the pointer **pIn** is pointing, the pointer **pIn** is moved one position to the right. When a pointer comes to the end of the array, it has to be moved to the beginning of the array.

A **FIFO** is also called a circular buffer.

The different PLC types frequently offers a **FIFO** in their PLC software library. It is very useable; it is, however, often not possible to adjust it (it is often locked with a password), if it does not satisfy the needs, and then the possibility to move the PLC code to another PLC type is limited. Therefore, on the next pages, it is shown that the implementation of a **FIFO** can be used and adjusted for free to your requirement.

Below is shown an example of a solution:

```
FUNCTION_BLOCK FIFO
VAR_INPUT
    // Insert data into buffer
    DataIn : REAL;
    // 0 : Do nothing, 1 : Insert data, 2 : Take of data
    INOutStatus : INT;
END_VAR
VAR_OUTPUT
    // Take out data from the buffer
    DataOut : REAL;
END_VAR
VAR CONSTANT
    // Max fixed size of the buffer
    BufferMax : INT := 5;
    // Min fixed size of the buffer
    BufferMin: INT := 1;
END_VAR
VAR
// Current no of data points (elements)
 NoOfDataPoints : INT := 0;
// Array having all elements
Buffer: ARRAY[BufferMin..BufferMax] OF REAL;
//Pointer to first element
 pIn : INT := 1;
//Pointer to last element
pOut : INT := 1;
END_VAR
```

In order to make the PLC code transparent, the function block has been added a control variable, named **INOutStatus**. This variable can have three status settings: If the variable is 0, nothing is done in the function block. If the variable is 1, the value, which exists on the **DataIn**, must be inserted in the queue. If **INOutStatus** is 2, a value must be taken out of the queue and be placed in the **DataOut** variable.

Two BOOL variables could have existed instead of **InOutStatus**, as a BOOL would be easier to use in a LADDER program.

```
///////////////////////////////////////////////////////////////////////////////////
// FIFO - First In First out
// Can handle up to BufferMax REAL data points
// If more REAL data points entered, the old one will be overwritten
///////////////////////////////////////////////////////////////////////////////////

//Insert data into buffer
IF INOutStatus = 1 THEN
  IF pIn <= BufferMax THEN
    Buffer[pIn] := DataIn; //Insert
    //Increase number of data points
    IF NoOfDataPoints < BufferMax THEN
      NoOfDataPoints:= NoOfDataPoints + 1;
    END_IF
    pIn:= pIn + 1; //Set to next element
  ELSE // buffer full, insert into first element
    pIn:= BufferMin;
    Buffer[pIn] := DataIn;
    //Move pointer to next element
    pIn:= pIn + 1;
  END_IF;
END_IF;

//Take out data of the buffer
IF INOutStatus = 2 THEN
  IF NoOfDataPoints > 0 THEN //There must be data
    Dataout:= Buffer[pOut];
    Buffer[pOut] := 0; //Set to 0 to show that the value is removed
    NoOfDataPoints:= NoOfDataPoints - 1;
    IF pOut < BufferMax THEN
      pOut:= pOut + 1;
    ELSE
      pOut:= BufferMin;
    END_IF;
  END_IF;
END_IF;

//Is buffer full? Last value is overwritten, move pIn pointer
IF NoOfDataPoints >= BufferMax THEN
  pIn := pOut;
END_IF;
```

```
PROGRAM MAIN
VAR
    OutData: REAL;
    MyFIFO: FIFO;
END_VAR
```

```
//Insert 71 and 35 into the FIFO
MyFIFO (DataIn:= 71, INOutStatus:= 1);
MyFIFO (DataIn:= 35, INOutStatus:= 1);

//Take out the first inserted value
MyFIFO (INOutStatus := 2 , DataOut => OutData);
//OutData = 71
```

13.3 Generating random numbers (RND, Randomize)

This chapter shows how a few lines of PLC code is able to generate random numbers. The random numbers can be used for testing a PLC control, where numbers can e.g. be the weight or the size of a subject which must be packed. In this way, the PCL control can be tested with many different numbers – a test which is very close to a test with real subjects.

Very often no access is given to real production subjects in order to test the PLC control, so by simulating values with a generator, as shown below, it is possible to test as much PLC coding as possible at the office, before the commission test.

By testing the PLC code early in the phases of development, possible programming faults and bugs are found and corrected. Later they are more difficult to find.

The PLC code is written in a **FUNCTION_BLOCK** named **RND**:

```
FUNCTION_BLOCK RND
VAR_INPUT
    Seed: INT;  // Start value, a value below ValueMax
    ValueMax: INT; // Max value to be generated
END_VAR
VAR_OUTPUT
    ValueRandom: INT; // The returned randomized value
END_VAR
VAR
    RandomSeed: DINT := 0;
END_VAR
```

```
//////////////////////////////////////////////////////////////////////////////////////////////////////
// This function is a randomize function
//
// The function generates a different number each time the function is called
// The seed value set the start value and this can be taken from the PLC
// main clock time to ensure different start numbers
//
// Refer to: "The C Programming Language," by Kernighan and Ritchie:
//
// INPUT: Valuemax is the max value ( + / - ) of the range
// INPUT: Seed, start just a number below max
// OUTPUT: ValueRandom a number in the range - ValueMin and ValueMax
IF RandomSeed = 0 THEN //Init
  RandomSeed := Seed;
END_IF
RandomSeed := RandomSeed * 1103515245 + 12345;
ValueRandom := DINT_TO_INT((RandomSeed / 65536) MOD (ValueMax + 1));
```

The **RND** function block is used as follows:

```
PROGRAM MAIN
VAR
  MyRND:   RND;
  NewValue: INT;
END_VAR
```

```
MyRND (Seed:=5, ValueMax:=10, ValueRandom => NewValue);
```

As shown above, the variable **MyRND** is created in the **MAIN** program with the data type **RND** and a variable **NewValue** is created to contain the random number.

With the above values, **NewValue** will become a number between -10 and 10 after each program execution. When all numbers between -10 and 10 have been 'drawn out', it all is repeated from the beginning again. Notice that the numbers occur in the same order and the distribution is mathematically evenly spread out in the whole interval -10 to 10. Having the same start value to **Seed**, numbers occur in the same sequence. **Seed** can advantageously be taken from the built-in clock in the PLC to secure different start values and hereby also secure more random numbers.

13.4 Digital Low-Pass filter (LP-Filter)

This chapter shows the implementation of a digital low pass filter. This filter is based upon a **Low Pass** (LP-Filter), consisting of an electronic coil in serial connection with an electronic capacitor (RC-filter). This filter lets the low frequencies pass and remove the high frequencies and is well-suited to remove noise signals.

On all the analogue input module, an LP filter is normally built-in, so that it is possible to filter noise and unwanted deflections from sensors and measuring equipment. Normally, it is not possible to modify the filter frequency online on an analogue input card and in some cases a modification of the frequency online is needed.

The example shown below is a 1st order digital filter, which is also called an exponential filter.

A Fourier transformation (advanced math) is used for transferring the analogue filter to a digital filter.

Many kinds of filters exist for Digital Signal Processing (DSP) and among the most well-known is a FIR (Finite Impulse Response).The advantage by using a digital filter instead of an average of data as e.g. 'moving average' is that 'moving average' includes all values and uses a long **ARRAY** for this. A digital filter removes the outside values and is fast for a PLC to work with.

A **FUNCTION_BLOCK** is used, as the filter must use a value from the previous program scan and this value is saved in **ValueOld**.

```
FUNCTION_BLOCK LP_Filter
VAR_INPUT
    ValueRaw : REAL; // Input value
END_VAR
VAR_OUTPUT
    ValueFiltered : REAL; // The filteret output value
END_VAR
VAR
    k : REAL; // Filter constant
    ValueOld : REAL;
END_VAR
```

```
//////////////////////////////////////////////////////////////////////////
//First order lag filter (LP-Filter)
//////////////////////////////////////////////////////////////////////////
//Versions log
//19.05.2018 TOAN, Created

k := 0.01; //Filter constant value

ValueFiltered := k * ValueRaw + (1 - k ) * ValueOld;

ValueOld:= ValueFiltered;
```

The filter frequency is adjusted by modifying the filter constant k:

k > 0.01	The filter is fast and does not remove a lot of signal.
k = 1	The filter is not working. (filter turned off)
k < 0.01	A lot of signal is filtered out (cutoff), and the signal takes a long time to come into the right signal level.

It is the PLC scan time, which is the sampling time. In practice, **k** must be adjusted, so that signal receives the nice curve which is wanted.

In the next chapter, find a PLC code example.

13.5 Simulation signals

This chapter describes simulation signals, which can be used during development and subsequent tests. The machine or the control hardware is often not available, when the PLC code is written and subsequently tested. The hardware has possibly not arrived, the machine is not fully constructed yet or the equipment is already sent to the customer. Therefore, it can be advantageous to be able to simulate 'sensor' signals on digital or analogue inputs and see if the PLC program is working as expected.

Below are four suggestions to simulation signals are shown, being easily adjusted in frequency and amplitude.

Signals can be connected in order to create new simulation signals:

```
MySignalCurve:= TriangleCurve + SinusCurve;
```

SINUS CURVE

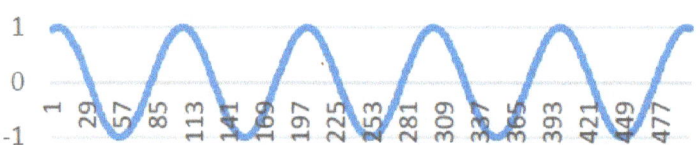

```
//This code generates a sinus curve
i:= i + 1; //Count to get a new value
IF i > 25 THEN
    i:= 1;
    n:= n + 1;
END_IF
SinusCurve:= SIN (n * 0.1); //0.1 to set  Hz
```

SQUARE WAVE / RECTANGLE CURVE

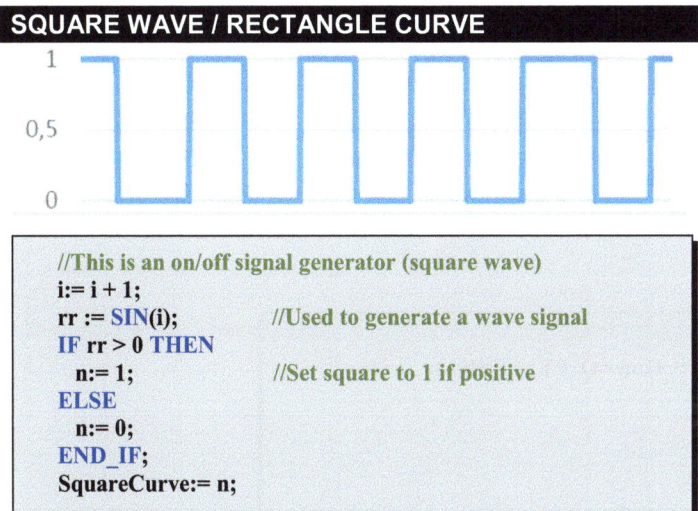

```
//This is an on/off signal generator (square wave)
i:= i + 1;
rr := SIN(i);          //Used to generate a wave signal
IF rr > 0 THEN
   n:= 1;              //Set square to 1 if positive
ELSE
   n:= 0;
END_IF;
SquareCurve:= n;
```

SQUARE WAVE / RECTANGLE CURVE (FILTERED)

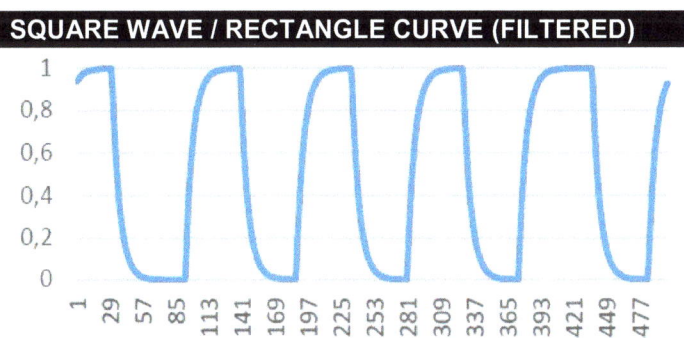

This signal is a squared/rectangle curve with filter.

A digital low pass filter is used on a square wave / rectangle curve signal.
For the **LP_Filter**, see chapter 13.4, page 100

```
LP_Filter ( ValueRaw:=  SquareCurve,
            ValueFiltered=> Filtered);
```

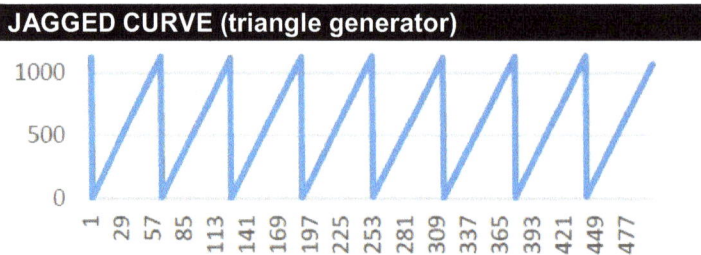

```
//This code generate a triangle curve
MyTimer(IN := NOT  MyTimer.Q, PT := T#10S); //Auto reset

//Timer end, go to zero
IF MyTimer.Q = TRUE THEN
   TriangleValue := 0;
END_IF;

//Add more and more to the curve (1.1321 is setting the slope)
TriangleCurve:= TriangleCurve + 1.1321;
```

Noise Signal

It is possible to add noise or signal deflections to the simulation signals by using a random generator (See chapter 13.3, page 98) and add the generated noise value to the signal:

```
MySignalCurve:= TriangleCurve + SinusCurve + NoiseSignal;
```

Data Plot

The shown graphs are plotted in Excel. First of all, the values are saved as ASCII log files (CSV files) on the hard disk by a soft PLC. Afterwards, the files are load and plotted in Excel.

13.6 Calculating tank volume, cylinder on hemisphere

This chapter shows an implementation of a volume calculation for a large storage tank.

The tank consists of a cylinder together with a hemisphere in the bottom.

Formulas for calculation of volume are found on the internet.

A FUNCTION is created where the size of the tank is input values, so that the PLC code can be reused for tanks of other sizes. Furthermore, the liquid height is input parameter to the function and the return value is the current volume. The liquid height is measured by an analogue sensor. It can be a pressure sensor in the bottom of the tank, measuring the liquid height or a sensor in the top measuring from the top downwards to the liquid height. The content in the tank is often the determining factor for which sensor technology to be used. In this solution example, the level is measured from the bottom of the tank up to the liquid height.

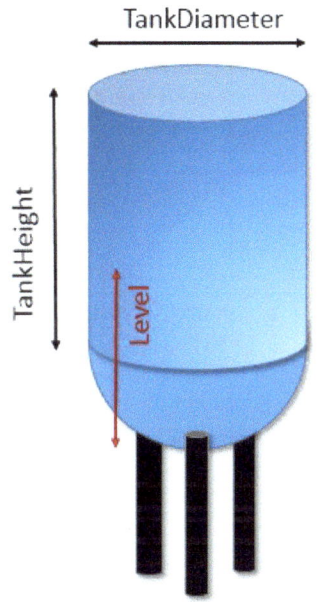

The mode of operation in this task is first of all to make the measurement succeed. Firstly, the liquid is poured into the cylinder, when the hemisphere is filled. The measurements must be checked by a tank calculator downloaded from the internet.

The function is chosen to be without units in order to make a more flexible solution, which can be reused and it means that all units must be equal. Units can be measured in meter, ft, cm or mm. The volume becomes cubic: m^3, ft^3, cm^3 or mm^3.

Next, the calculation of the hemisphere must work and the whole solution must finally be connected, so that a total test can be made. It is an advantage to document the test in a document to prove that the function is tested. To have a valid test, a range of test points must be selected, which must be placed outside the measure range of the tank and at different levels in the tank and close to the interfaces, being where the cylinder and the hemisphere meet. Firstly, the expected volume is calculated at different levels. It can be made by a calculator or by one of many online pages on the internet, where it is possible to make volume calculations on tanks. Finally, the function is tested including the different levels and the result is compared with the expected results.

Below a suggestion to the PLC code is shown:

```
FUNCTION TankVolumenCal : REAL
VAR_INPUT
    TankDiameter:      REAL;  // Fixed tank diameter
    TankHeight:        REAL;  // Fixed tank height of cylinder
    LevelFromButtom:   REAL;  // Current level measured
END_VAR
VAR CONSTANT
    PI: REAL := 3.1415;
END_VAR
VAR
    Level: REAL;        // Internal calculation
    Vol:   REAL := 0;   // Internal calculation
    Lr:    REAL;        // Level radius in circle
    TankRadius: REAL;
END_VAR
```

The program is split up in different clear sections as shown on the next page. In the first two lines the internal variables are initialized. Next, the calculation sections occur, where each section provides a comment line for information, and finally the return value for the function is set.

Program 'call' for the function could be as follows:

```
Vol1:= TankVolumenCal (TankDiameter:= 2,
                       TankHeight:= 6,
                       LevelFromButtom:= LevelSensor);
```

Or like this, because it is a FUNCTION:

```
Vol1:= TankVolumenCal (2, 6, LevelSensor);
```

Where **LevelSensor** is the current tank measurement.

All values must have the same unit (m, mm, cm, feet, ft).

```
/////////////////////////////////////////////////////////////////////////////////////////////
//  Tank Volumen calculator - Cylinder with a half circle below
/////////////////////////////////////////////////////////////////////////////////////////////
Level:= LevelFromButtom;
TankRadius:= TankDiameter/2;

//Check level low - level cannot be negative
IF Level < 0 THEN
  Level:= 0;
END_IF;

//Check level high - tank cannot be overfilled
IF Level > (TankRadius + TankHeight) THEN
  Level:= TankRadius + TankHeight;
END_IF

//Half circle ball
IF Level <= TankRadius THEN
  Lr:= SQRT(Level * (TankDiameter - Level));
  Vol:= (PI/6)*level*(3*Lr*Lr+ Level*Level);
ELSE
  //Half circle ball filled
  Vol:= 2/3*PI * TankRadius * TankRadius  * TankRadius;
END_IF;

//Something in the cylinder
IF Level > TankRadius THEN
  Vol:= Vol + (Level - TankRadius) * PI * TankRadius * TankRadius;
END_IF;

 //Set return value
TankVolumenCal:= Vol;
```

14 From LADDER to ST-programming

This chapter contains a range of examples, comparing LADDER programming with the corresponding programming in ST.

This chapter is meant to support the readers who understand LADDER programming well or where a LADDER program must be translated to ST. No converting tools exist, which are able to convert a LADDER program into an ST program, which is the reason for the following examples:

Example 1:

```
//Solution 1A
K1:= S1;

//Solution 1B
IF S1 = TRUE THEN
   K1:= TRUE;
ELSE
   K1:= FALSE;
END_IF;
```

Example 2:

Example 2A

```
//Solution 2
VAR
   S1_TRIG: R_TRIG;
END_VAR

S1_TRIG (CLK:= S1);
IF S1_TRIG.Q = TRUE THEN
   K1:= TRUE;
ELSE
   K1:= FALSE;
END_IF;
```

Example 3:

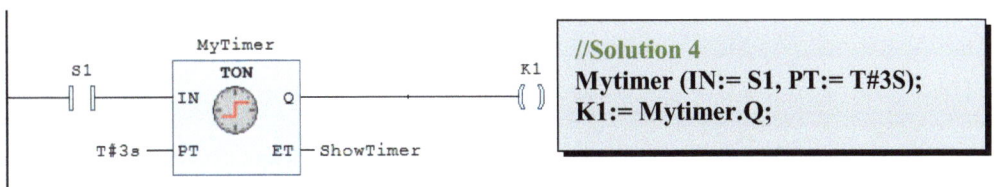

```
//Solution 3A
K1:= (S1 OR S2) AND S3;

//Solution 3B
IF ((S1 = TRUE OR S2 = TRUE) AND S3 = TRUE) THEN
   K1:= TRUE;
ELSE
   K1:= FALSE;
END_IF;
```

Example 4:

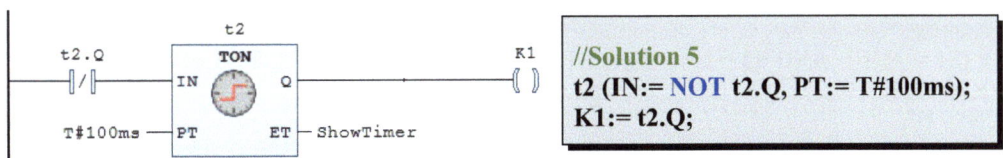

```
//Solution 4
Mytimer (IN:= S1, PT:= T#3S);
K1:= Mytimer.Q;
```

Example 5:

```
//Solution 5
t2 (IN:= NOT t2.Q, PT:= T#100ms);
K1:= t2.Q;
```

Example 6:

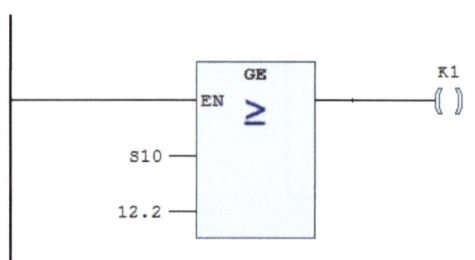

```
//Solution 6
IF S1 = TRUE THEN
  K1:= TRUE;
END_IF;

IF S2 = TRUE THEN
  K1:= FALSE;
END_IF;
```

Example 7:

```
//Solution 7A
K1:= S10 >= 12.2;

//Solution 7B
IF S10 >= 12.2 THEN
  K1:= TRUE;
ELSE
  K1:= FALSE;
END_IF;
```

Example 8:

```
//Solution 8A
IF (S1 = TRUE AND S2 = FALSE) THEN
  K2:= 123;
END_IF;
```

```
//Solution 8B
IF S1 AND NOT S2 THEN
  K2:= 123;
END_IF;
```

Example 9:

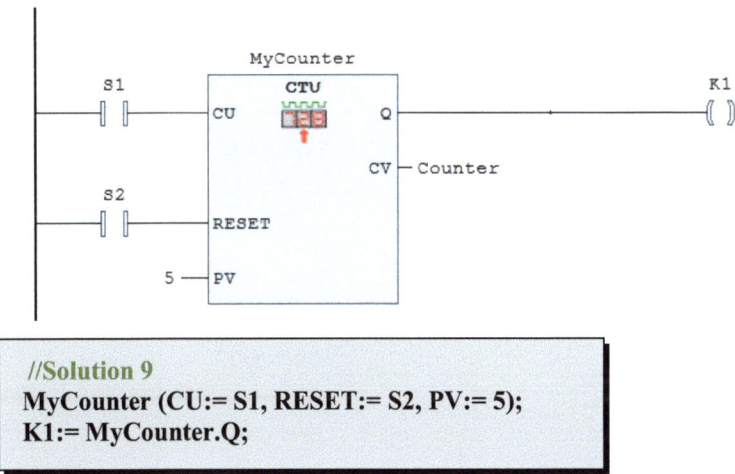

//Solution 9
MyCounter (CU:= S1, RESET:= S2, PV:= 5);
K1:= MyCounter.Q;

Example 10:

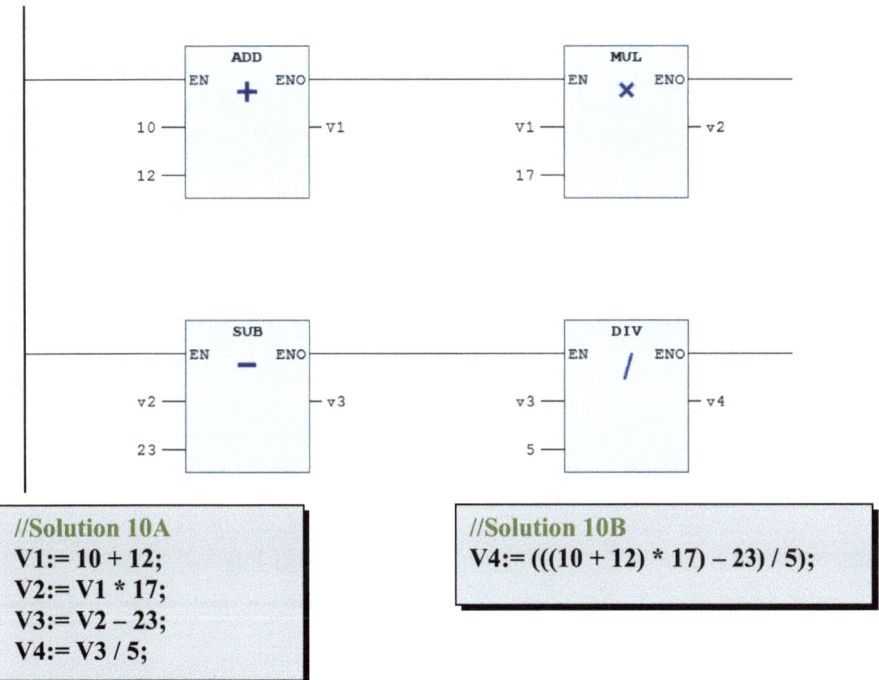

//Solution 10A
V1:= 10 + 12;
V2:= V1 * 17;
V3:= V2 – 23;
V4:= V3 / 5;

//Solution 10B
V4:= (((10 + 12) * 17) – 23) / 5);

Example 11:

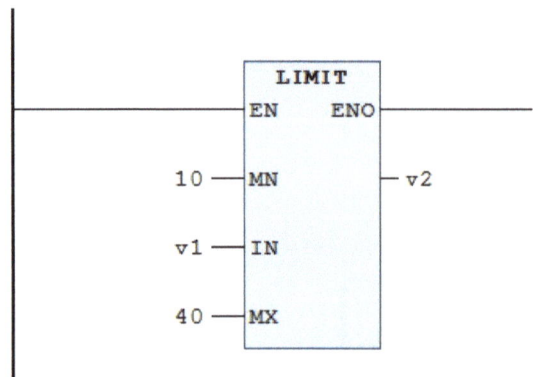

```
//Solution 11
v2:= v1;
IF v2 < 10 THEN
   v2:= 10;
END_IF;

IF v2 > 40 THEN
   v2:= 40;
END_IF;
```

Example 12:

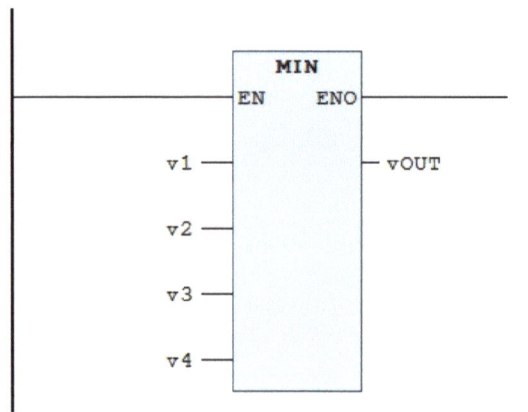

```
//Solution 12
vOUT:= v1;

IF vOUT > v2 THEN
   vOUT:= v2;
END_IF;

IF vOUT > v3 THEN
   vOUT:= v3;
END_IF;

IF vOUT > v4 THEN
   vOUT:= v4;
END_IF;
```

Example 13:

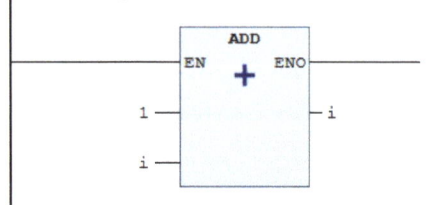

```
//Solution 13
i:= i + 1;
```

15 Best Practice ST-programming

Even if ST programs make it possible for each programmer to use his/her own typographs (syntax), a good practice within programming have to be followed in order to increase the readability in the entire program. Capital and small letter typographs together with tabulation/space can improve the readability of the program.
It is important to write the program in a uniform setting, so that other readers can easily read it.

Below, a summary and recommendations are shown:

15.1 Indentation and SPACE

Tabulation is typically relevant when using **IF** and **CASE** statements and **FOR** loops. The best solution is to use 2 x <SPACE> for tabulation, as <TAB> is depending on the setup in the PLC developing tool and Windows settings. If the PLC code must be copied later to another PLC, the best solution is 2 x <SPACE>.

Tabulation increases the readability of the PLC code. On the other hand, the PLC code can be difficult to read without tabulation or by making a wrong tabulation. Consequently, the recommendation is to use the same tabulation in the entire program.

<SPACE> has no function in ST programming; create, however, one <SPACE> between commands, variables, statements, brackets and values, because such a <SPACE> increases the readability of the code. It is, however, recommended not to place <SPACE> before semicolon.

15.2 Empty line between code

It makes sense to have empty lines in the PLC code in order to separate and split up the different code pieces in suitable parts.

It is, however, recommended to make two empty lines max between PLC codes.

15.3 Avoid spaghetti code

The spaghetti code is a designation for PLC code possessing a complex and complicated structure which occurs when unclear naming of variables and functions, many **GOTO**s, **JMP**s, **EXIT**s or other unstructured implementations.

It is, however, recommended only to use **GOTO** and **JMP** statements in very special situations (e.g. for fault finding, test and debug). Furthermore, the use of **EXIT** can cause a spaghetti code. It is, therefore, recommended to avoid the **EXIT** command and use other conditioned statements as **IF** and **CASE**. On the other hand, **EXIT** can be useful, when a fault finding is carried on a PLC code; care must, however, be taken when the PLC code is finalized and remember to remove the **EXIT**s, which are not used any longer.

It is, however, allowed to use **EXIT** in FOR-loops, if it is not necessary to execute the whole loop. For implementation of **EXIT** in a FOR-loop see chapter 9.4, page 62

15.4 Use functions and program modules

The most basic way of building a manageable structure is by using program modules and functions. By splitting up a large program into more smaller programs, each with a specific task, it is possible to create a small main program (MAIN), 'calling' sub programs (program modules) when needed.

Functions and function blocks are effective as the PLC code can easily be reused and when correcting it, only one place has to be corrected.

It is also recommended to give the functions and modules an indicative name, so that they are easy to recognize.

If the function or a program provides more than 20 local variables, it is an indication that the PLC code must be split up in more functions or program modules.

15.5 Use of variables

Very often it must be decided whether local or global variables are to be used. The use of global variables is fast and easy as they are only to be declared once in a common variable/TAGS list. It creates, however, a bad program structure and the variables because all functions and program modules have access to them.

It is recommendable to use local variables if possible, and delete the variables which are not to be used anymore. It is all about not having many variables, but use the ones with an indicative name.

It is recommendable to use **STRUCT** to collect the variables in an object.

It is recommendable only to create **ARRAY** with the needed length.

If a function or a program module provides more than 20 local variables, it could be an indication of a bad structure and the program ought to be split up even more.

Perhaps some PLC code can be moved to a function and reused.

15.6 Miscellaneous

Below, a couple of other programming hints are stated:

- Exchange complicated **IF-THEN** statements with a **CASE** statement
- Avoid **ELSIF** statements
- Avoid infinite loops, and consequently **DO-WHILE** are not recommended
- Do not use more than 3 incorporated loops in FOR loops
- Each function/module must provide max. 20-25 line codes – as what you see on the screen when programming and on a paper print out.
- Do not use more than three-dimensional arrays (3D **ARRAY**)
- Use **CONSTANT** if the same constant number is used more than once
- Program modules or functions must provide max. 20 local variables

Avoid of creating too many unnecessary **ARRAY** elements. They are easy to create, and unfortunately, some programmers prefers creating too many and too long elements, which leads to too many system resources.

It is recommended to use parentheses in math formulas and algorithms to make sure that the sequence of the calculation is correct.

15.7 Code sharing with the internet

Google has a unique value for a programmer to find code on the internet. The largest problem is to find useful PLC code. Sometimes it takes a long time to find useful code, because it might include faults. The consequence of this makes it easier for you to write your code yourself. Copyright on the code can be the reason why you cannot use it, if you or your company earns money by using what is found on the internet.

Another challenge when finding a code on the internet is the fact that names on variables and structure are often not corresponding to the standard and naming which are chosen in your own program. Often names are not corrected and overall more time is spent on finding code, making fault corrections and adjusting code, rather than using time on writing the code yourself. Care must be taken not to upload your own code on the internet, if you are employed in a company, because the code is the company's property and it can be considered as theft.

Examine as well your company's policy concerning codes including comments and discussions of other companies' PLC codes and programming solutions on the internet so that no problems occur later. It can go against 'The Employers' and Salaried Employees' Act' in Denmark and elsewhere, because you contribute a yield for others, which not directly creates a value for the company where you are employed.

15.8 OOP – Object-Oriented Programming

In order to structure the PLC code better, the philosophy from Object-Oriented-Programming (OOP) can be used. This means that variables and PLC codes, which are related to each other, are collected in an object. Variables which are e.g. used for a motor are collected in a **STRUCT** (see chapter 4.3, page 19) and the operational conditions for a motor are collected in **ENUM** (see chapter 4.4, page 21)

Variables and constants which e.g. work on the same **ARRAY** can have the same first name in order to mark their relationship, so that they work on an object, a component, an instrument or one specific task at the time.

Some PLC types offer OOP, as described in the standard IEC 61131-3. These PLC types offer **METHOD** (mode of operation as a function), **ACTION** (mode of operation as a program module) and **PROPERTY** and **TRANSITION**.

16 Guide to programming exercises

This chapter is a guide which can help the reader, when solving programming exercises.

1) Get started

Read the task, and as a rule read it more than once. It is important only to solve exactly what the task describes and nothing more than that, as it is often a customer is to use the solution and he/she will not pay extra. If more is solved than what the task describes, the program will create more faults, which are often experienced as a bad quality.

If the task is not well described, it is important to examine any uncertainties which had occurred. It can be described in a document creating a total overview of how the control has to work. The document is called a function description or control description. A well-functioning document is written to retain knowledge and to be shown to the customer.

2) I/O-list

Work out an I/O list. Study what the individual sensors and instruments have to measure and how they work. The I/O list is an important tool both during the development of the control, commissioning and the continuous maintenance and possible expansion. It is important that the I/O list is more than 95 % percent correct before starting the programming, because changes in the I/O list have certain influences on the programming and the subsequent test.

Make indicative and reasonable names for variables/TAGS already in the I/O list, because the names are common for the whole project and the I/O list is a part of the documentation. If the task, diagram or document already consists of indicative and reasonable names, these are used to make sure that they are identifiable.

3) HMI

Most control solutions contain a user manual, which consists of HMI (frames) and perhaps electrical on/off contacts and lamps. Make a rough outline suggestion on paper/hard copy of how frames could look. Show the suggestions to the customers/users or a colleague to gain feedback. It is time consuming to correct pictures later. Therefore, it is important that the pictures are as correct as possible, before starting the configuration of the HMI.

It might be advantageous to work out a list of which variables/TAGS to be exchanged between HMI and the PLC program as an interface description always gives a good overview. It might possibly not be the same persons who are coding the HMI and the PLC. This is the reason why a list is perfect guidance for both persons.

4) Flowcharts

Work out flowcharts for the complex program parts, so that you have a better feeling of how the control must work. Flowcharts are good guidance for you and others who need to understand the program and how it works.

5) Design Phase

Before startup the programming, it might be advantageous to work out a design draft on paper, which contains the different program modules, functions and function blocks. It could be like flowcharts and it is possible to use flowcharts as the description and can be seen as a program design phase. This description also determines the names of the program modules and functions shortly describing each program module and function. A certain experience is needed to be able to design a complete program before starting of the programming, and therefore it is advantageous to use the bottoms-up method, as described in the next part of this chapter.

6) The programming

There are 2 possibilities when starting up (implementation) the programming. It is the top-down method or the bottoms-up method.

Bottoms-up is defined by writing PLC code to the small programming parts, which you know must be used. You start, so to speak, by writing what is clear to you. If e.g. the control has a lamp which has to flash, a piece of PLC code, which can flash, is written. Gradually a range of small well-run PLC codes are collected – small building blocks. Gradually a lot of knowledge is gained and more and more you get the feeling of how the entire program is functioning.

Finally, it becomes easier to compose the entire program from the small pieces. HMI can advantageously be used gradually to test the small pieces to make sure that they before they together make an entire program. It becomes more difficult to make the entire program, if many of the small pieces do not work correctly and fault finding in a large program is more difficult than in the small program pieces. Tests of small programs are often called module tests and how they are tested can advantageously be documented via e.g. screen dumps of the current program, so that you can document to yourself and others that it works well.

It might be a help to work on two projects (in the PLC developing tools) at the same time. A project becomes the final solution and a project is going a test (a sand box) of different small program pieces. Small solutions in a project are tested and when the solution works well, it is copied (copy pasted) (or the code is rewritten in order to have a nice looking structure) into the final project.

A PLC developing tool, executed in a Windows environment, can crack up by a Run-time Error (or blue screen of death) and therefore it is a good idea often to save the PLC code, which must be done every time the PLC code is working well. Then you can always go back, if any troubles should occur and the project file is destroyed.

If you are in doubt of how individual small programs can be implemented, then use Google to be inspired. There are many solutions on the internet which can give a lot of inspiration. Sometimes more time is spent searching the internet rather than trying to code on your own. Remember, if any problems occur, do not use more than e.g. 15 minutes before you go on with other tasks and ask your company's support department, google or ask a colleague about the problem in order to use your time the best possible way. Often small things during programming are ignored – or are missing in the manual and you cannot solve your problem within 15 minutes, then you cannot even solve it within 60 minutes.

17   Subject Index